Science

New Jersey ASK
Science Test Preparation

Grade 6

Visit *The Learning Site!*
www.harcourtschool.com

Printed in the United States of America

ISBN 0-15-353505-9

1 2 3 4 5 6 7 8 9 10 018 15 14 13 12 11 10 09 08 07 06

Contents

Learning About New Jersey ASK Science Test Preparation

The next two pages give examples of the types of items found on the science portion of the Assessment of Skills and Knowledge (ASK) and Grade Eight Proficiency Assessment (GEPA).

About Multiple-Choice Items

Many of the items on the Assessment of Skills and Knowledge (ASK) and Grade Eight Proficiency Assessment (GEPA) are multiple-choice. Each multiple-choice item has four answer choices. The tips that follow will help you answer these questions.

1. Read the question carefully. Restate the question in your own words.

2. Watch for key words such as *best*, *most*, *least*, or *except*.

3. The question might include tables, graphs, diagrams, or pictures. Study these carefully before choosing an answer.

4. Find the best answer for the question. Fill in the answer bubble for that answer. Do not make any stray marks around answer spaces.

1. Which simple machine is a wedge?

Tip
Think about the characteristics of a wedge.

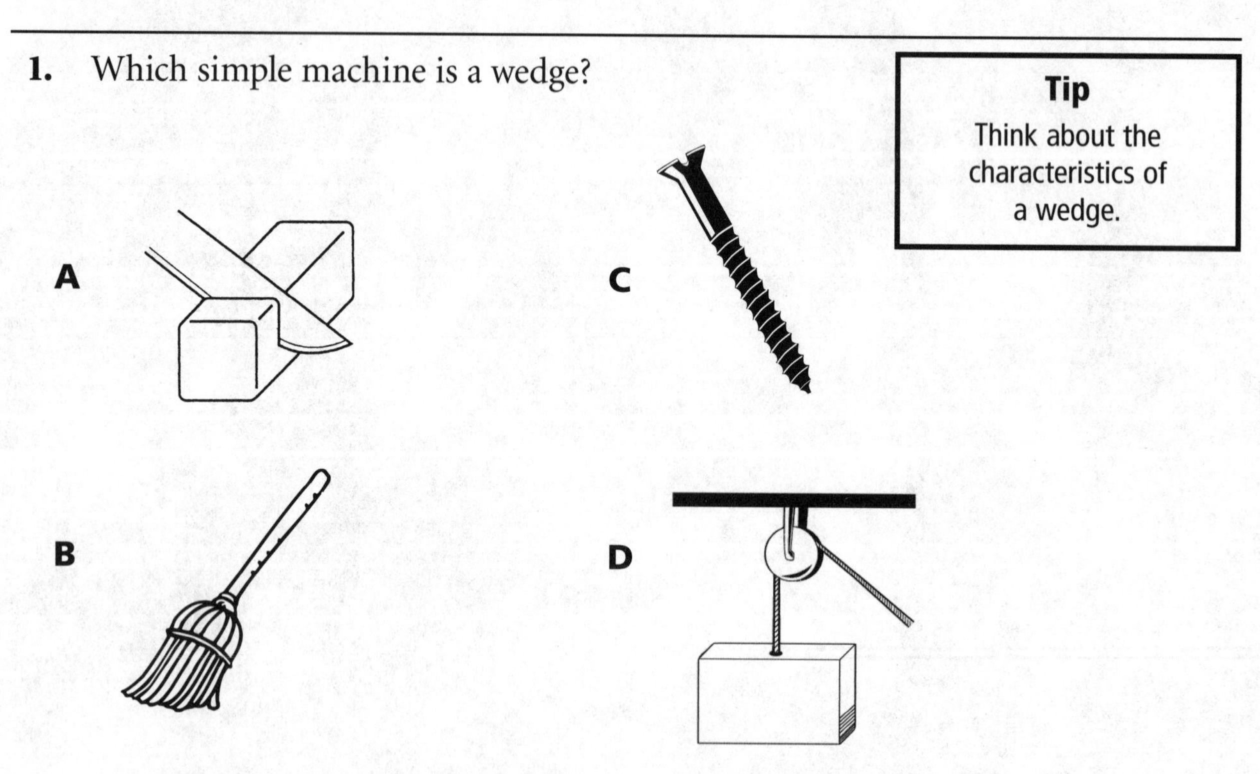

Learning About New Jersey Science Assessments

About Constructed-Response Items

For some items, you must write a brief answer to explain a science concept or to apply a science process skill. To receive the highest score answers should

- be complete.
- show understanding of the science content and processes.
- be accurate.
- communicate the ideas clearly.

---

1. Study the stage of mitosis shown in the diagram.

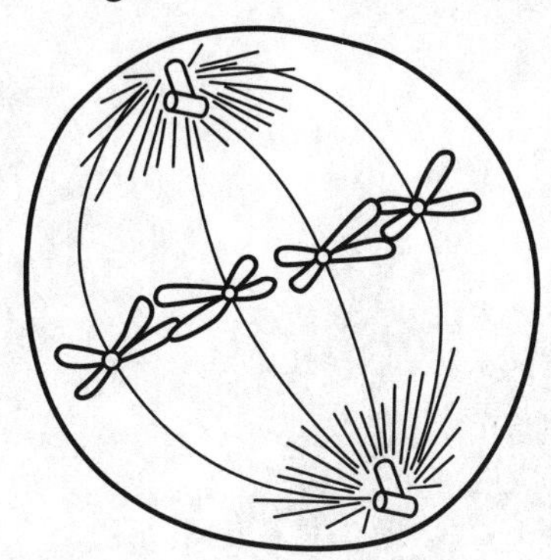

> **Tip**
> Study the position of the chromosomes in the diagram.

What is happening during the stage shown? What will happen next?

____

____

____

____

About Practice Sets

Each practice set consists of 3 multiple-choice items, 1 two-point short-answer item, and 1 four-point short-answer item. Some items have Tips, which give clues about how to answer the items. An item may have a graph, table, picture, or diagram. Study these carefully before answering the items.

Name ______________________________

Date ______________________________

Tip

Think about the inquiry skills used to interpret a graph.

1. When Ramon and his family moved into their new town, they noticed that many families had dogs as pets. Ramon began recording each type of dog he saw. He made a graph of his data.

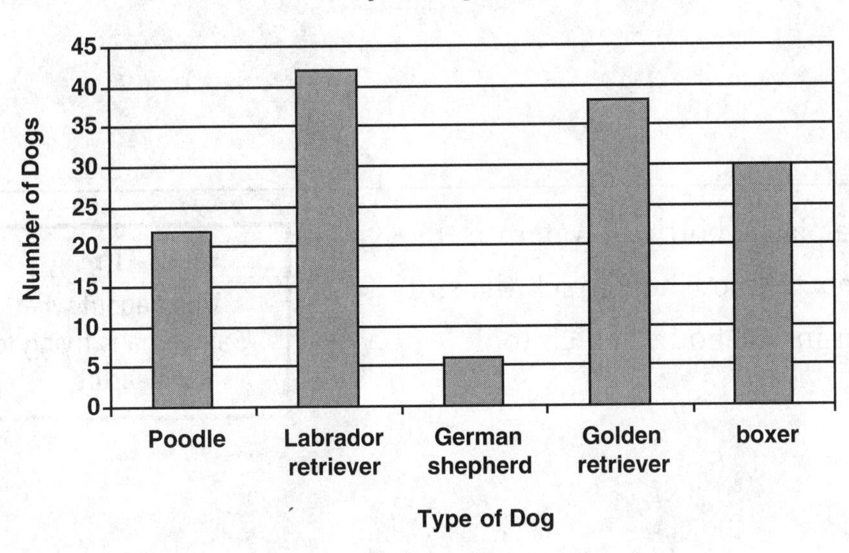

Which of these inquiry skills is Ramon **least likely** to use to determine which type of dog is the most popular in his new town?

A comparing

B using numbers

C interpreting data

D measuring

Name ______________________ Date ______________________

2. When conducting an investigation, which inquiry skill uses your senses to gather information?

A classify/order

B compare

C predict

D observe

---

3. Samantha places a plastic bottle of water on the sidewalk. She wants to know how much the sun will heat the water in one hour. Which tool should she use?

> **Tip**
> Think about what Samantha is trying to measure.

A 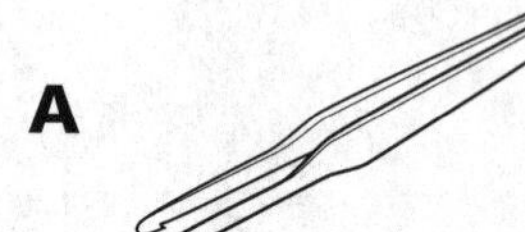

C 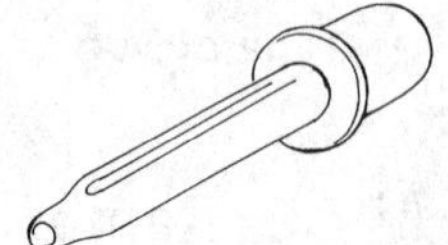

B 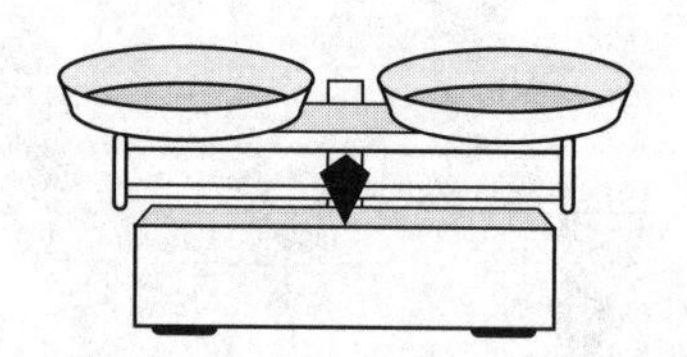

D 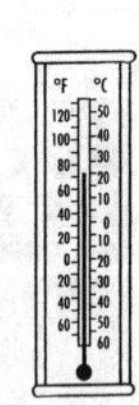

Name ______________________ Date ______________________

4. You want to find out which brand of plastic wrap works best. You hypothesize that Brand A will keep foods freshest. You design an investigation to test your hypothesis. What inquiry skill(s) will you use to determine which plastic wrap is most effective?

____

____

____

Tip

Think about the inquiry skills you can use in an investigation. Which would be most useful when evaluating the quality of plastic wrap?

Name ____________________ Date ____________________

5. Your teacher has given you the following object and instructed you to measure it.

> **Tip**
> Think about the shape of the ball. Remember that you can measure more than just an object's size.

• What property of the soccer ball could you measure?

____

____

____

• What tool would you use to make this measurement?

____

____

____

Name ______________________________

Date ______________________________

1. Study the animal cell below.

> **Tip**
> Study the labels. Recall the function of each cell part.

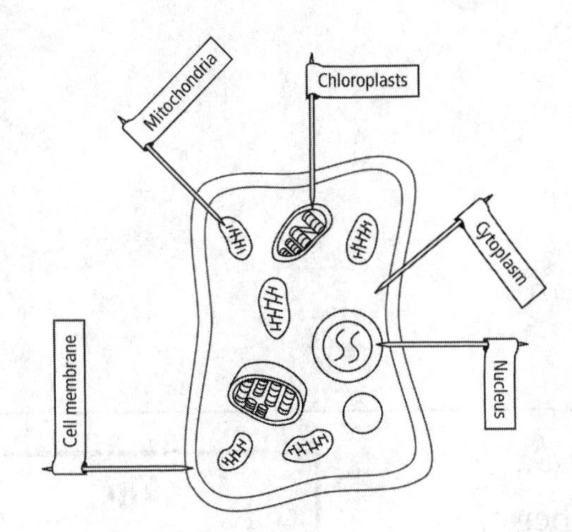

Which structure performs respiration?

A nucleus

B mitochondria

C vacuole

D chloroplast

Name ______________________ Date ______________________

2. Which sequence best describes the organization of an organism made up of many cells?

A tissue → organ → organ system → cell

B cell → tissue → organ system → organ

C tissue → cell → organ → organ system

D cell → tissue → organ → organ system

---

3. How does the number of chromosomes in a human body cell compare with the number of chromosomes in a human reproductive cell?

A There are the same number in a body cell as there are in a reproductive cell.

B There are twice as many chromosomes in a body cell than in a reproductive cell.

C There are twice as many chromosomes in a reproductive cell than in a body cell.

D There are half as many chromosomes in a body cell than in a reproductive cell.

Tip

Compare the way in which chromosomes are divided in mitosis with the way they are divided in meiosis.

Name ______________________ Date ______________________

4. Study the stage of mitosis shown in the diagram.

Tip
Study the position of the chromosomes in the diagram.

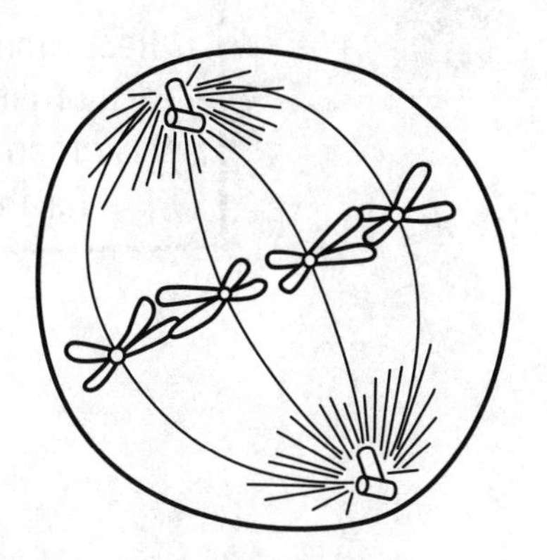

What is happening during the stage shown? What will happen next?

____

____

____

____

____

Name ______________________ Date ______________________

5. Study the diagram of the human organs.

> **Tip**
> Recall the function of each organ and the job this organ system does for the body.

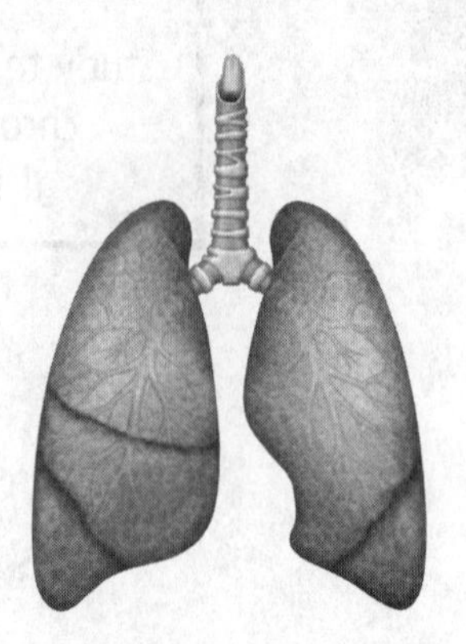

• Name each organ, and tell which organ system they are part of.

____

____

____

• Explain how the organs work together.

____

____

____

____

____

Name ______________________

Date ______________________

Tip
Remember the characteristics of the cells making up the organisms in each of the five kingdoms.

1. Study the organisms.

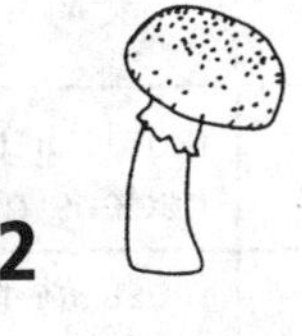

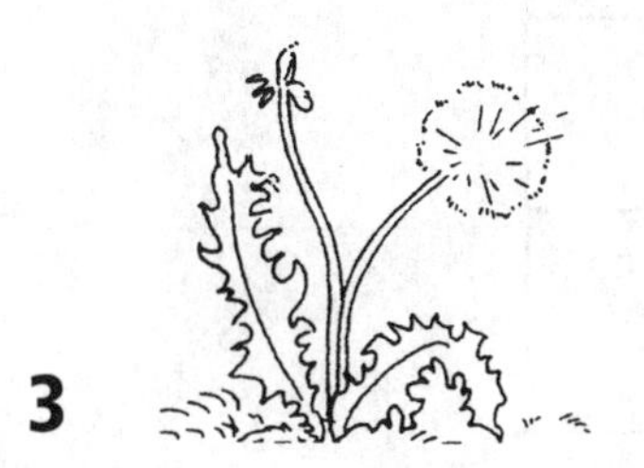

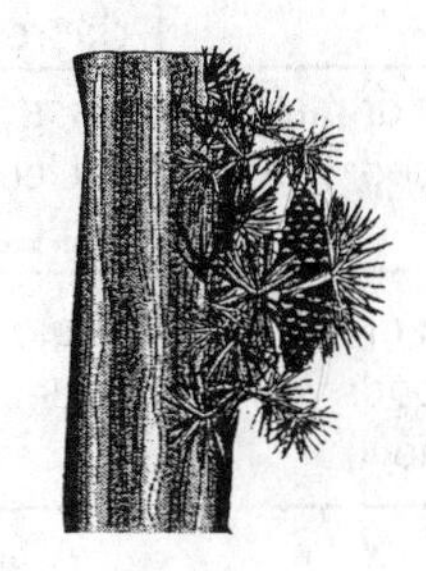

Which organisms can be grouped into the same kingdom?

A 1, 2, and 3

B 1, 2, 3, and 4

C 1, 2, and 4

D 1, 3 and 4

---

2. How do scientists classify organisms?

A by their internal and/or external features

B according to where they live

C by their size and color

D by their appearance

Name ______________________ Date ______________________

3. What are the features of the cells of members of the kingdom Plantae?

> **Tip**
> What do plant cells have that animal cells do not have?

The Kingdoms of Living Things		
Kingdom	**Description**	**Features of Cells**
Plantae	make their own food; cannot move	?
Animalia	eat food; most can move	multicelled, no cell walls, nucleus, no chloroplasts
Fungi	take in food; cannot move	most are multicelled, cell walls, nucleus, no chloroplasts
Protists	take in food or make their own food; most can move	mostly single celled, nucleus, some have cell walls and chloroplasts
Bacteria/Monera	take in food or make their own food; some can move	single celled, cell walls, no nucleus

A no cell walls, chloroplasts, nucleus

B cell walls, nucleus, chloroplasts

C cell walls, no nucleus, no chloroplasts

D no cell walls, no nucleus, chloroplasts

Name ______________________ Date ______________________

4. Look at the illustration of the animal.

> **Tip**
> Recall that the two parts of the classification that are used in the scientific name.

The animal belongs to the Kingdom Animalia, phylum Chordata, class Mammalia, order Carnivora, family Felidae, genus Felis, and species Domesticus. What is the correct way to write the name for this animal? By what name is this animal commonly called?

____

____

____

Name ______________________ Date ______________________

5. Before a class field trip, your teacher provides you and your team with a dichotomous key.

Tip
Make sure to use all necessary information when you use a dichotomous key.

Dichotomous Key		
1.	a. solid-red head	Red-headed woodpecker
	b. red cap on head	go to 2
2.	a. black-and-brown wings	common flicker
	b. black-and-white wings	go to 3
3.	a. white belly.	Hairy woodpecker
	b. tan belly	Red-bellied woodpecker

- Describe how you and your classmates will use the key. What features will you use to identify a bird?

____

____

____

____

____

- When you are in the field, you see a bird that has a red cap, black and white wings, and a tan belly. What kind of bird is it? How can you tell?

____

____

____

____

____

Name ______________________________

Date ______________________________

Tip
Recall the characteristics of monocot and dicot plants.

1. Look at the plant in the figure below. How many cotyledons do the seeds in the plant most likely have?

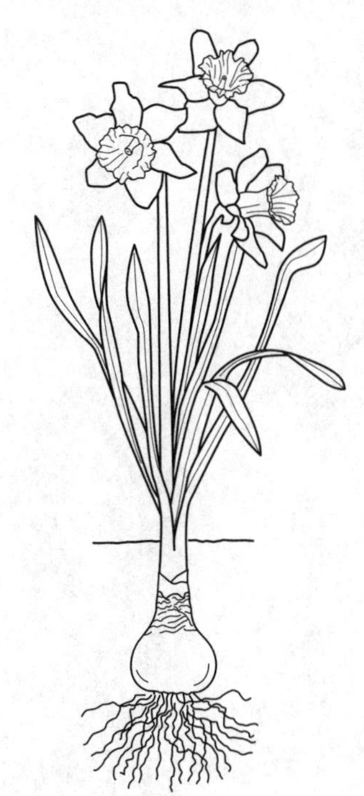

A one

B two

C three

D four

Name ______________________ Date ______________________

2. The following is a cross section of the vascular tissue in a plant root.

Tip
Think about the materials a plant's vascular tissues carry.

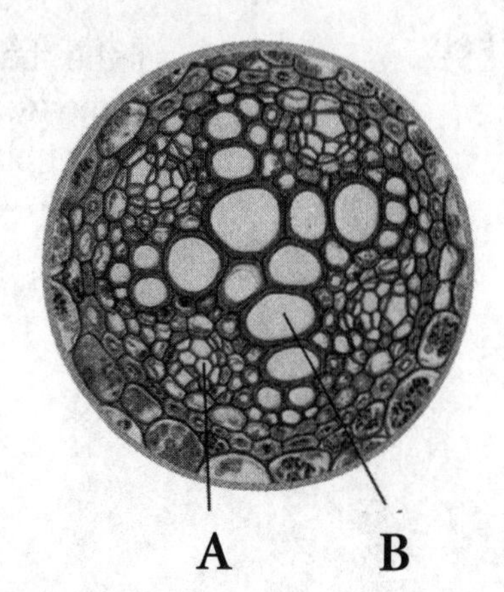

What is the function of the tissue labeled B?

A attracting pollinators

B transporting sugar

C transporting water and nutrients

D converting sunlight to sugar

---

3. Which method listed is a form of sexual reproduction?

A runners

B seeds

C leaf cuttings

D tubers

Name ______________________ Date ______________________

4. The illustration shows part of a plant.

Tip
Remember the characteristics of mosses, ferns, gymnosperms, and angiosperms.

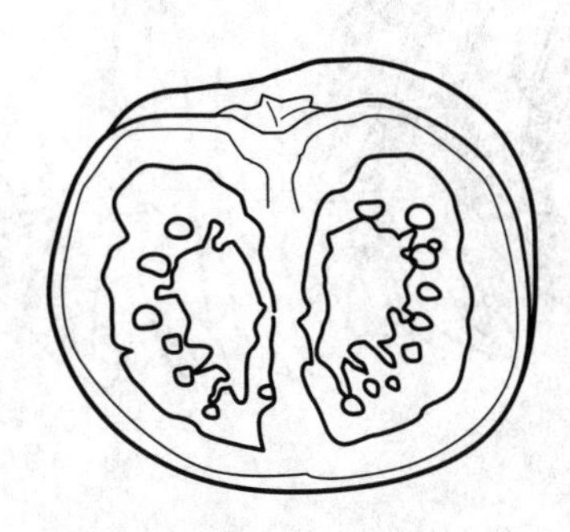

How would the plants from which these plant parts came be classified? How can you tell?

____

____

____

____

____

Name ______________________ Date ______________________

5. When on a hike, you find a plant.

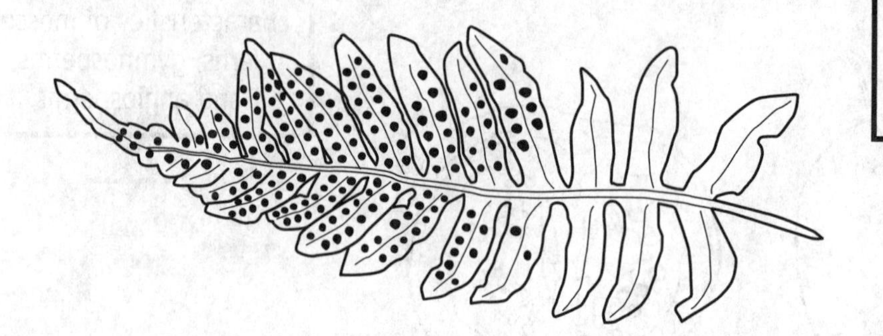

> **Tip**
> Focus on both the physical characteristics of the plant and the way it reproduces.

You observe the plant, then refer to the chart.

Plant	Description	Type of Reproduction
A	small plants; no vascular tissue; one or two cells thick	asexual reproduction; spores
B	vascular plants; fronds; fiddleheads	two generations in life cycle; spores
C	vascular; most have needlelike leaves; many have cones	seeds not surrounded by fruit
D	vascular, flowering plants	seeds surrounded by fruit

- Use your observations of the plant, and the chart to classify it into one of the groups from A–D. Into which group should it be placed? What characteristics guide this classification?

____

____

- What kind of plant have you identified? How can you tell?

____

____

Name ______________________________

Date ______________________________

Tip
Think about the things that help organisms survive in the desert.

1. The picture shows several organisms in their environment.

Which is a structural adaptation to its environment in one of the organism's shown?

A the spines on the cactus

B the dormancy of plants

C the lizard's ability to warm its body

D the sandy soil

Name ______________________ Date ______________________

2. Which of the following shows a correct order in a grassland food chain?

A hawk → grass → snake → mouse

B grass → mouse → snake → hawk

C mouse → snake → grass → hawk

D mouse → hawk → snake → grass

---

3. Which of the following is a behavioral adaptation caribou have made to live in the tundra biome?

A estivation

B thick fur

C migration

D tropism

Tip

Think about the characteristics of the tundra to help you determine what behavioral adaptation helps caribou survive there.

Name ______________________ Date ______________________

4. The forest ecosystem has both living factors and nonliving factors.

Tip
Remember the difference between *biotic* and *abiotic*.

What abiotic factors are shown? What biotic factors are shown?

____

____

____

Name ________________________ Date ________________________

5. The graph shows the relationship between jackrabbit and coyote populations.

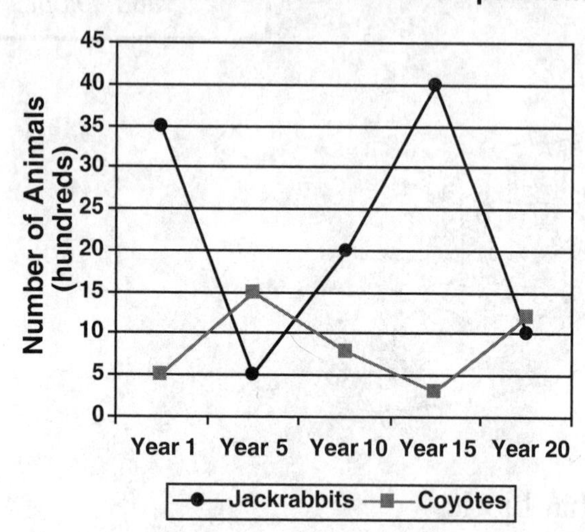

Tip

Recall the different feeding relationships that occur in nature.

• Describe the interaction between jackrabbits and coyotes. What type of relationship is shown?

____

____

____

____

• What pattern in population can you identify?

____

____

____

____

____

Name ______________________

Date ______________________

Tip
Recall examples of both renewable and nonrenewable resources you see or use every day.

1. Which item is a renewable resource?

A oil

B tree

C coal

D can

Name ______________________ Date ______________________

2. Which step is part of the carbon cycle?

> Tip
> Eliminate choices you know are **not** steps in the carbon cycle.

A Rain falls from clouds to Earth's surface.

B Animals release oxygen and take in carbon dioxide.

C Plants take in carbon dioxide and release oxygen.

D Bacteria in soil break down nitrogen compounds.

---

3. Which environmental condition threatens the wildlife of the Florida Everglades?

A natural destruction of habitats by wildfire

B human destruction of forests for building

C human destruction of grassland to build farms

D human control of water that used to flow freely

Name ______________________ Date ______________________

4. The Northern spotted owl is threatened due to human destruction of its habitat.

> **Tip**
> Recall the definitions of threatened species and endangered species.

What can you infer about the future of the Northern spotted owl? How can humans affect it?

____

____

____

____

____

Name ______________________ Date ______________________

5. During a hike through a national park, you spot an unusual plant. The park service has posted a sign stating that it is unlawful to pick the plant because it is endangered.

Tip
Think about the meaning of the term *endangered*. How do human activities affect endangered populations?

• What is an endangered species?

____

____

____

____

• You want to have a specimen of the plant for your science notebook. You think, "How can picking just one flower hurt?" What should you do, and why?

____

____

____

____

____

____

____

____

Name ______________________________

Date ______________________________

Tip
Picture the layers of Earth from the center to the surface.

1. Which is the correct order of Earth's layers from the center to the outer surface?

A inner core → mantle → outer core → crust

B crust → mantle → outer core → inner core

C inner core → outer core → mantle → crust

D lithosphere → inner core → asthenosphere → mantle

---

2. How is energy released when an earthquake occurs?

A as lava

B in the formation of a tsunami

C as heat

D in the form of waves

Name ______________________ Date ______________________

3. Study the boundary shown in the diagram.

> **Tip**
> Define each of the answer choices, and then decide which definition matches the figure.

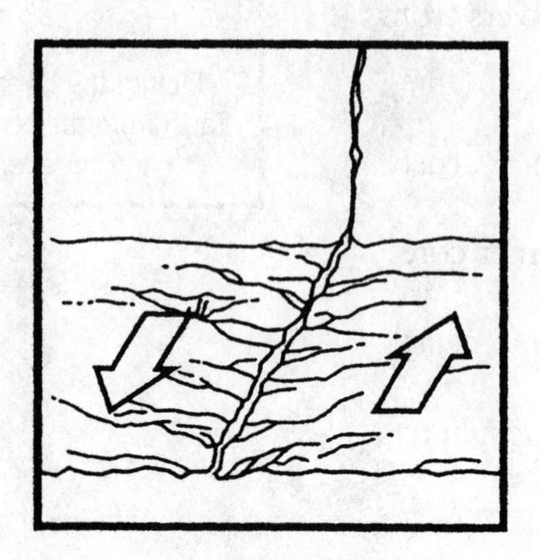

What kind of plate boundary is shown?

A divergent boundary

B mid-ocean ridge

C convergent boundary

D transform fault boundary

Name ______________________ Date ______________________

4. The diagram shows earthquake strength of four earthquakes.

> **Tip**
> Remember that the Richter scale assigns a number value to an earthquake.

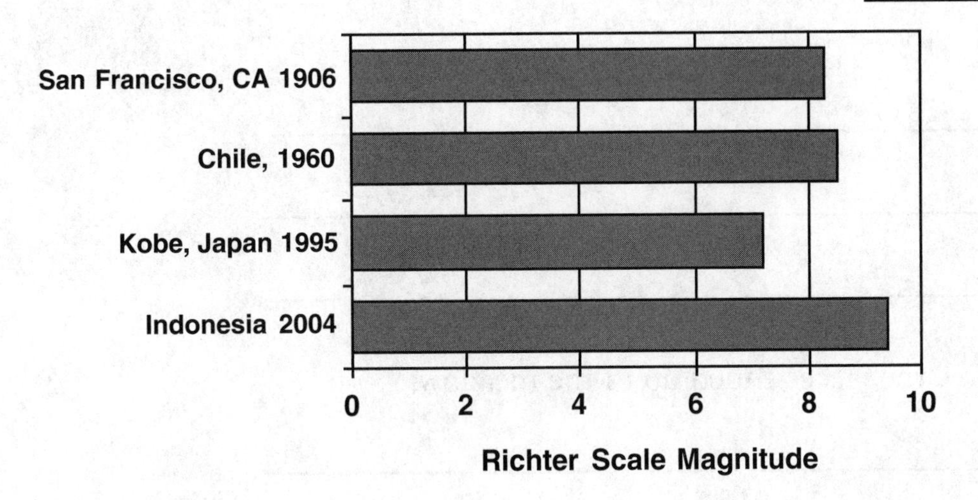

Explain approximately how much stronger the San Francisco earthquake of 1906 was compared to the Kobe, Japan earthquake of 1995.

____

____

____

____

____

____

Name ______________________ Date ______________________

Tip
Review the different weathering forces that change Earth's surface.

5. A large boulder once sat atop a hill. Now, it is small pieces in a meadow below.

- Describe the types of weathering forces that might have acted on the boulder while it was still on top of the hill.

____

____

____

- Explain how the pieces ended up in the meadow.

____

____

____

Name ____________________

Date ____________________

Tip
Try to recall the term that matches each definition, and eliminate the incorrect answers.

1. Scientists use many different properties to identify minerals. What are they observing when they observe a mineral's cleavage?

   A the color of the powder that is left behind when a mineral is rubbed on a piece of unglazed tile

   B the shape of the crystals that make up the mineral

   C the way a mineral breaks along a flat surface parallel to the crystal face

   D the measure of a mineral to resist being scratched

---

2. Which describes the formation of igneous rocks?

   A Molten rock cools, hardens and crystallizes.

   B Pieces of rocks and minerals become cemented together.

   C Rocks are exposed to an increase in temperature and pressure.

   D Water evaporates inside a sea shell and crystals form on the inside.

Name ______________________ Date ______________________

3. Which of the following is a method that may increase erosion?

Tip
Think about the factors that cause erosion.

A planting plants with deep roots

B contour plowing

C planting windbreaks

D planting on steep hillsides

Name ______________________ Date ______________________

4. Look at the diagram of the rock cycle.

> **Tip**
> Recall the different forces acting on the different types of rocks.

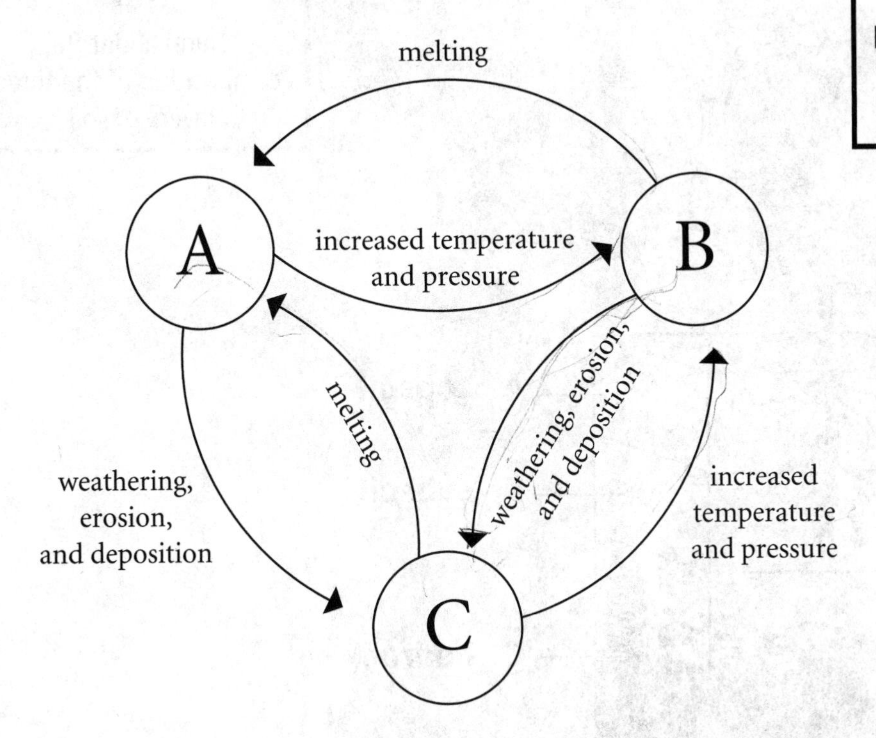

Explain how an igneous rock might become a sedimentary rock.

____

____

____

____

____

Name ______________________ Date ______________________

5. The diagram shows the composition of soil.

> **Tip**
> Think about the composition of the three layers of soil.

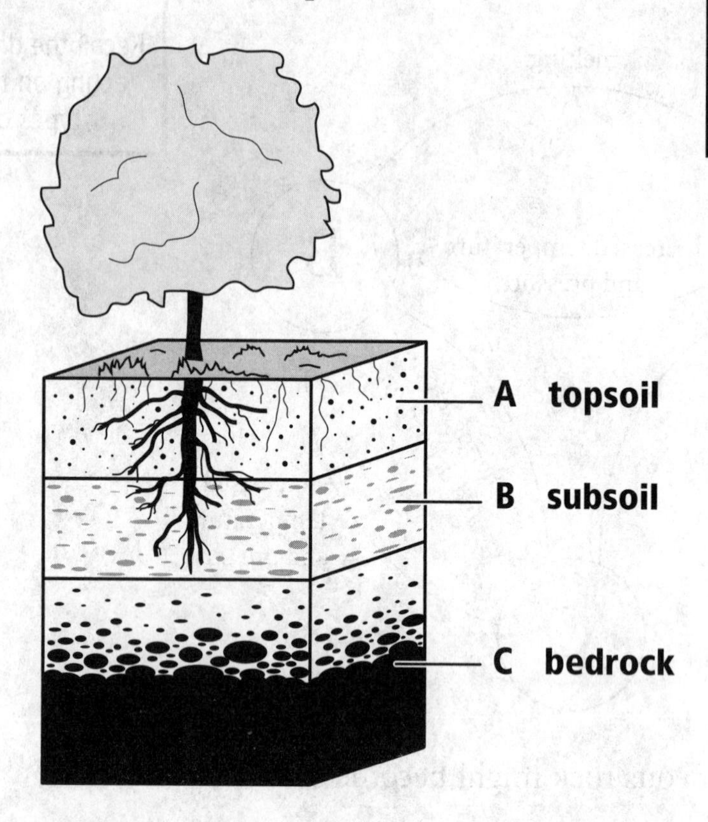

• Describe the composition of topsoil.

____

____

____

• Describe the composition of subsoil.

____

____

____

Name ______________________

Date ______________________

Tip
Remember that petrified wood is a fossil.

1. What causes petrified wood to form?

A Cells of wood are slowly filled with minerals.

B Sediment hardens around a piece of wood.

C A piece of wood is frozen for thousands of years.

D A piece of wood becomes trapped in tree sap.

---

2. What are index fossils?

A fossils of insects trapped in tree sap

B fossils of organisms that lived during a short time span

C fossils of organisms that lived over millions of years

D fossils that are found in only one location on Earth

Name ______________________ Date ______________________

3. Examine the chart below.

Tip
Recall the steps in coal formation.

Coal Formation
A. Moisture and oxygen are pressed out. → Swamp plants die and sink to the bottom of a swamp. → Decayed matter is buried under sediment. → Bacteria and fungi decompose plant matter.
B. Swamp plants die and sink to the bottom of a swamp. → Bacteria and fungi decompose plant matter. → Decayed matter is buried under sediment. → Moisture and oxygen are pressed out.
C. Bacteria and fungi decompose plant matter. → Decayed matter is buried under sediment. → Swamp plants die and sink to the bottom of a swamp. → Moisture and oxygen are pressed out.
D. Swamp plants die and sink to the bottom of a swamp. → Moisture and oxygen are pressed out. → Bacteria and fungi decompose plant matter. → Decayed matter is buried under sediment.

Which flow chart shows the correct order in the formation of coal?

A A

B B

C C

D D

Name ______________________ Date ______________________

4. Examine the list of fossils and where they were collected.

Type of Fossil	Environment Fossil Found In
land snail shell	in the dirt on the forest floor
coral skeleton	in the rocks beneath the forest floor
clam shell	sitting on the ocean floor
fish skeleton	lying on the river bottom

Tip

Recall that all living things have specific environmental requirements.

Which fossil indicates that the environment has changed over time?

____

____

____

____

Name ______________________ Date ______________________

5. The diagram shows several layers of soil and rock.

> **Tip**
> Think about how the layers of rock are formed.

Rock Layers

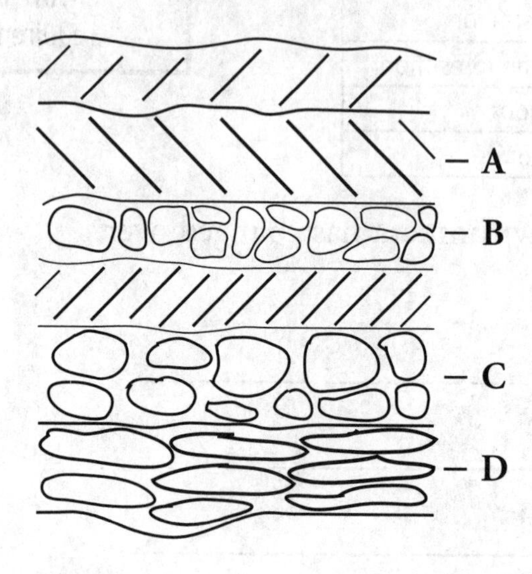

• Which layer of rock is the oldest? Explain your reasoning.

____

____

____

____

• Which layer of rock contains the oldest fossils? Explain your reasoning.

____

____

____

____

____

Name ______________________________

Date ______________________________

1. Which of the following is a source of salt water?

A glaciers

B polar ice caps

C inland lakes

D oceans

---

2. What is the correct sequence of events that take place at a water treatment plant to make water safe to drink?

> **Tip**
> Think about what needs to be done to wastewater to make it safe to drink.

A wastewater collected → sedimentation → primary treatment → oxygen reactors → settling → clean water

B primary treatment → wastewater collected → oxygen reactors → settling → sedimentation → clean water

C oxygen reactors → sedimentation → wastewater collected → primary treatment → settling → clean water

D wastewater collected → primary treatment → settling → sedimentation → oxygen reactors → clean water

Name ______________________ Date ______________________

3. The table lists the different zones of the ocean.

Zones of the Ocean	
1	intertidal zone
2	near-shore zone
3	open-ocean zone
4	abyssal zone

Tip
Think about which part of the ocean has continuous changes in water level.

In which zone must living things deal with an ever-changing environment?

A zone 1

B zone 2

C zone 3

D zone 4

Name ______________________ Date ______________________

4. The ocean floor has several different regions. Identify two features that are present on the abyssal plain.

Tip
Envision the characteristics of each part of the ocean floor.

____

____

____

____

Name ______________________ Date ______________________

5. Study the figure of the water cycle.

> **Tip**
> Recall the series of events that take place in the water cycle.

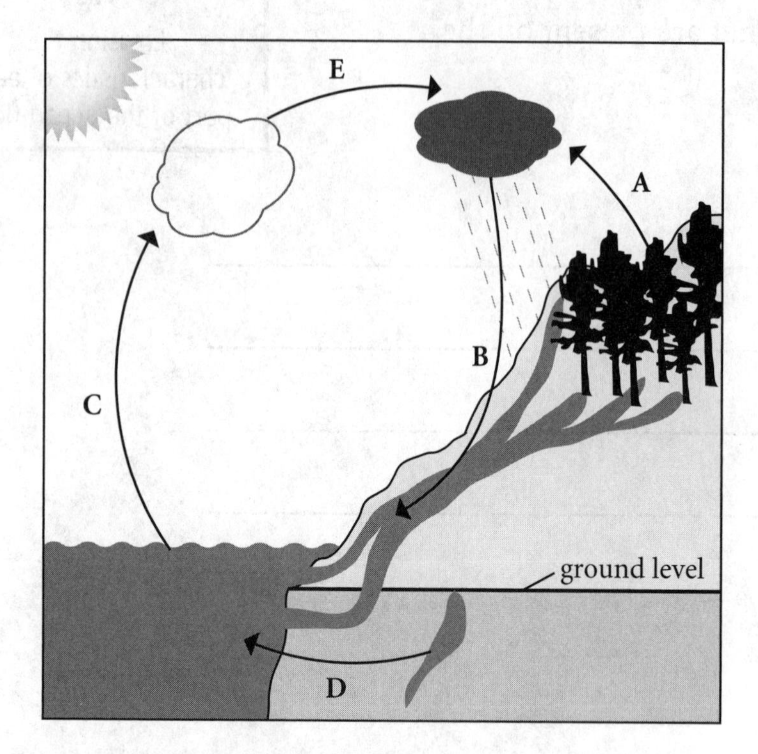

- Name and explain the process taking place at point C on the diagram.

____

____

____

____

- How does water return to the ocean in the water cycle?

____

____

____

____

Name ______________________________

Date ______________________________

Tip
Remember each layer of the atmosphere, and where it belongs.

1. The diagram shows the layers of Earth's atmosphere.

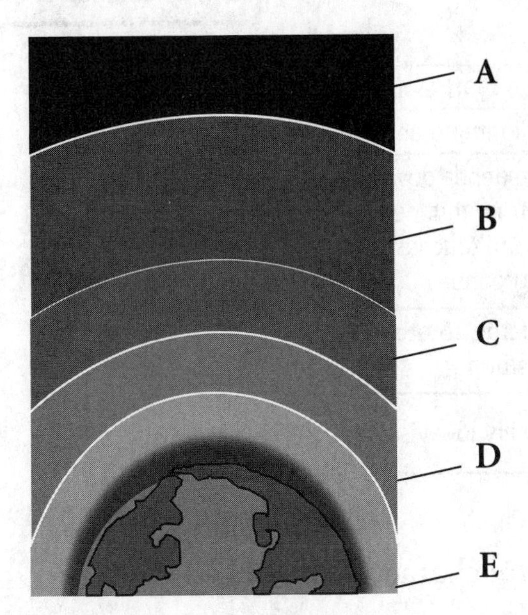

Which layer of the atmosphere is identified with the letter C?

A exosphere

B stratosphere

C thermosphere

D mesosphere

---

2. Which feature is often used to describe an air mass?

A humidity

B volume

C altitude

D density

Name ______________________ Date ______________________

3. Study the table of characteristics of storms. Each letter represents a type of storm.

Tip
Think about the characteristics of different kinds of storms.

	Wind	Precipitation	Other factors
A	strong and gusty	rain	lightning and thunder
B	violently rotating column of air	rain possible	extends downward from thunderclouds and touches the ground
C	wind speeds of at least 119 km/hr	heavy rain	large, rotating tropical storm
D	winds speeds at least 56 km/hr	heavy, blowing snow	very low visibility

Which type of storm has factors that are characteristic of a tornado?

A A

B B

C C

D D

Name ______________________ Date ______________________

4. What is relative humidity?

Tip

Use the definition of *humidity* to help you define *relative humidity*.

____

____

____

____

Name ______________________ Date ______________________

5. The diagram shows a cold front moving into an area.

> **Tip**
> Remember that warm air is less dense than cold air.

• Describe what happens to the air masses as the cold front moves in.

____

____

____

____

• What type of weather occurs when a cold front moves in?

____

____

____

____

Name ______________________

Date ______________________

1. What occurs during a solar eclipse?

A The sun's rays shining toward Earth create a shadow on the moon.

B The moon blocks the sun's rays from reaching Earth.

C The sun blocks the moon's rays from reaching Earth.

D The moon's rays shining toward Earth create a shadow on the sun.

---

2. The table contains information on the length of each planet's year.

Tip
Carefully read all of the information in the table.

Planet	Length of Planet's Year
A	88 Earth days
B	255 Earth days
C	165 Earth years
D	29.5 Earth years
E	88 Earth days
F	248 Earth years
G	84 Earth years
H	12 Earth years
I	365 Earth days

Based on this data, which planet is farthest from the sun?

A Planet C

B Planet F

C Planet B

D Planet I

Name ______________________ Date ______________________

3. In addition to the planets, other objects are found in the solar system. They are described in the table.

Tip
Recall the characteristics of objects other than planets that are found in the solar system.

Object	Description
asteroid	made up of rock and metal
meteor	rock smaller than an asteroid
comet	ball of ice, rock, and frozen gasses
satellite	bodies in space, such as moons

Which of these objects burns up in Earth's atmosphere?

A asteroid

B meteor

C comet

D satellite

Name ______________________ Date ______________________

4. Look at the diagram.

> **Tip**
> Recall how the seasons in the Northern and Southern hemispheres are different.

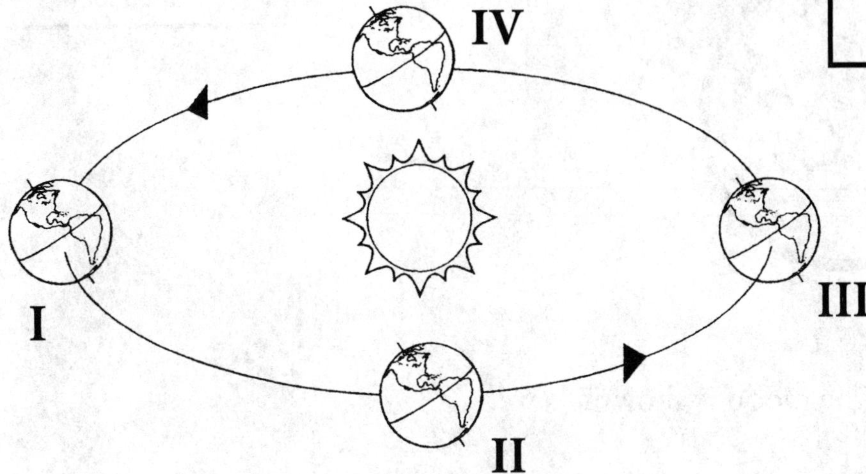

At position I on the diagram, explain what season each hemisphere is experiencing.

____

____

____

____

____

____

Name ______________________ Date ______________________

5. The picture shows one phase of the moon.

> **Tip**
> Remember that sunlight is always shining on the moon.

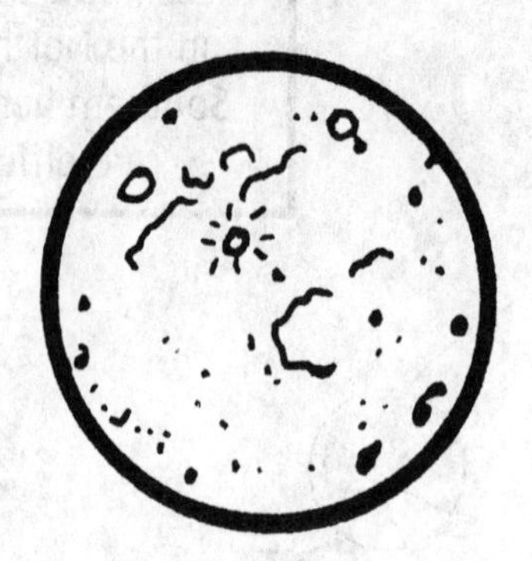

- What phase of the moon is shown?

____

____

- Explain how the sun, the moon, and Earth are arranged in order for this phase to be seen.

____

____

____

____

Name ______________________________

Date ______________________________

1. What is atomic number?

A the sum of the number of neutrons and electrons in an atom

B the difference between the number of protons and the number of neutrons

C the number of neutrons multiplied by the number of electrons

D the number of protons in a nucleus

---

2. The figure shows the box for the element oxygen taken from the periodic table.

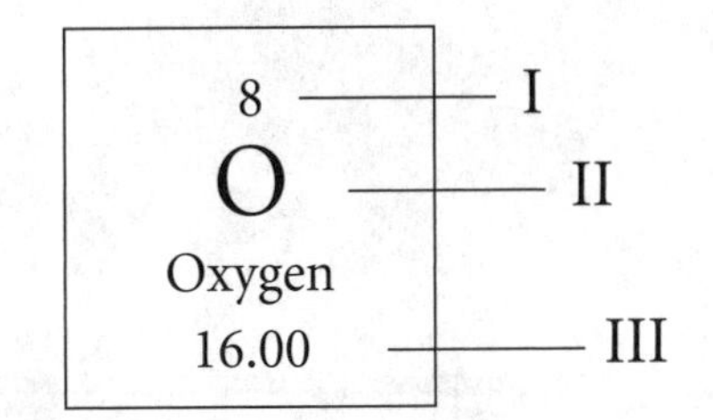

Tip
Think about how elements are arranged in the periodic table.

What is the number labeled III?

A atomic number

B atomic mass

C average atomic mass

D isotope number

Name ________________ Date ________________

3. Look at the part of the periodic table shown.

> **Tip**
> Recall the information given in the periodic table.

15	16	17	18
			2 He Helium
7 N Nitrogen	8 O Oxygen	9 F Fluorine	10 Ne Neon
15 P Phosphorus	16 S Sulfur	17 Cl Chlorine	18 Ar Argon
33 As Arsenic	34 Se Selenium	35 Br Bromine	36 Kr Krypton

What is the difference between nitrogen and oxygen?

A Nitrogen has one less proton.

B Nitrogen has one less neutron.

C Nitrogen has one less electron.

D Oxygen has one less proton.

Name ____________________ Date ____________________

4. The graph shows the boiling point of one cup of water.

> **Tip**
> Recall what you know about the boiling point of water and salt.

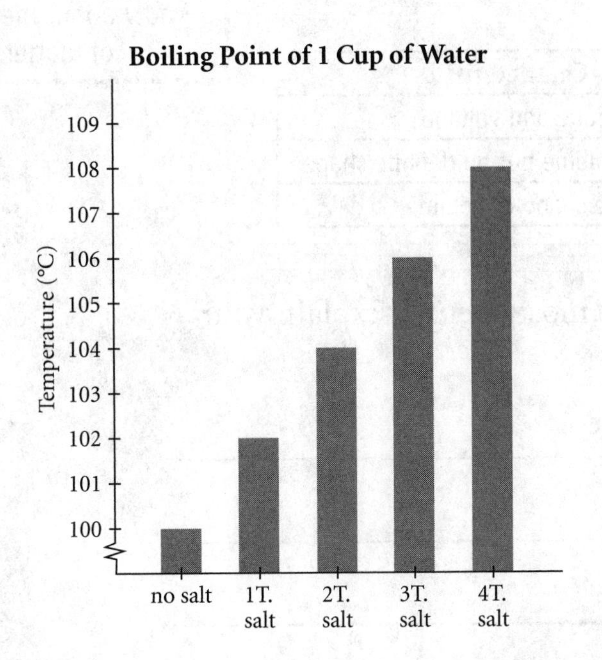

Summarize how the addition of salt affects water's boiling point. Why does this occur?

____

____

____

Name ____________________ Date ____________________

5. In nature, matter is found in different states, as described in the table.

Tip
Think about what you know about the states of matter.

State of Matter	Characteristics
solid	definite shape and volume
liquid	definite volume but no definite shape
gas	no definite shape or volume

- Which state of matter has the most energy? Explain your reasoning.

____

____

____

____

____

- What is required to change a solid into a gas?

____

____

____

____

____

Name ______________________

Date ______________________

1. Look at the tool used for scientific inquiry.

Tip
Recall the tools used for various measurements.

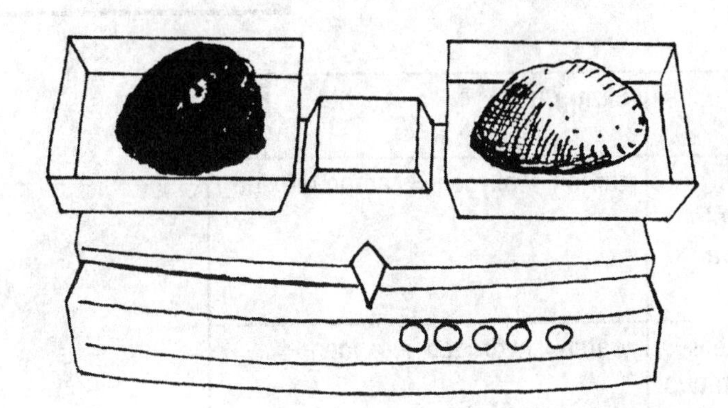

When you use this tool, which physical property of the rocks are you measuring?

A volume

B density

C temperature

D mass

2. Which of the following is a chemical change?

A water boiling

B grinding a board into sawdust

C burning the wick of a candle

D melting chocolate

Name ______________________ Date ______________________

3. Chemical reactions can be classified into four different types as shown in the table.

Tip
Use the table to review the four types of reactions.

Type of Reaction	Description	Example
synthesis	two or more substances are put together to form a single substance	carbon + oxygen = carbon dioxide
decomposition	a single substance is broken up to form two or more substances	water = hydrogen + oxygen
single replacement	one substance replaces another	magnesium + lead nitrate = lead + magnesium nitrate
double replacement	two substances switch places	sodium chloride + silver nitrate = sodium nitrate + silver chloride

Which type of reaction occurs when the brown gas nitrogen dioxide is heated, as shown in this equation?

hydrogen sulfide + silver = silver sulfide + hydrogen

A synthesis

B decomposition

C single replacement

D double replacement

Name ______________________ Date ______________________

4. The diagram shows the pH of some common substances.

Substance	pH
ammonia	11
vinegar	3
milk of magnesia	9
car battery acid	0
distilled water	7
drain cleaner	10
lemon drink	2

Tip
Use the table to remember how the pH scale represents acids and bases.

What substance is the strongest acid? How can you test whether a substance is an acid or a base?

____

____

____

____

Name ______________________ Date ______________________

5. Magnesium hydroxide is a medicine used to relieve indigestion caused by an excess of stomach acid.

> **Tip**
> Recall the relationship between acids and bases.

• How does magnesium hydroxide work?

____

____

____

____

• Could other substances be used to relieve indigestion? Why or why not?

____

____

____

____

Name ______________________________

Date ______________________________

Tip
Think about what happens to energy as a car rolls down a ramp.

1. Look at the drawing of the toy car on a ramp.

As the toy car sits at the top of the ramp, what kind of energy does the toy car have?

A potential energy

B kinetic energy

C thermal energy

D chemical energy

Name ______________________ Date ______________________

2. When fireworks explode, which energy transformation takes place?

A electrical energy to sound energy

B kinetic energy to light energy

C chemical energy to sound energy

D solar energy to potential energy

---

3. Which correctly states the law of conservation of energy?

Tip
Recall the meaning of the term *conservation*.

A The total amount of energy in a system can be increased in an emergency.

B The amount of potential energy in a system can be increased if the amount of kinetic energy is too low.

C The amount of energy in a system is always the same.

D Energy can be created, but it cannot be destroyed.

Name ____________________ Date ____________________

4. The diagram below shows a sound wave.

> **Tip**
> Recall the parts of a wave and how waves move.

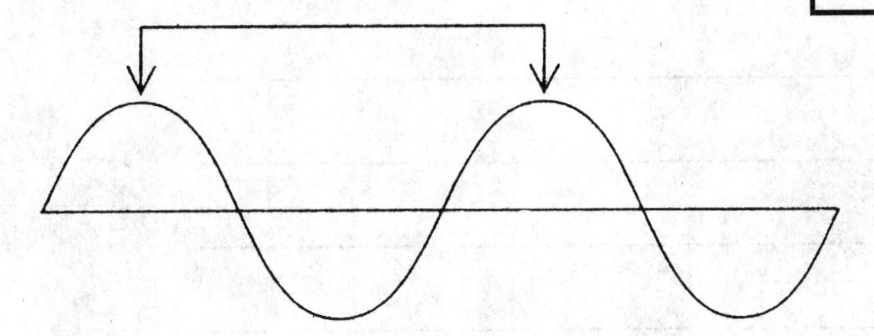

What part of a wave do the arrows in the drawing identify?
What would happen to sound if the amplitude were increased?

____

____

____

____

Name ____________________ Date ____________________

5. You turn on a CD player to listen to music.

Tip
Remember the law of conservation of energy.

• Describe the forms of energy you observe as you do this.

____

____

____

____

• What energy transformations take place when you turn on the CD player?

____

____

____

____

Name ______________________________

Date ______________________________

Tip
Think about how a circuit works.

1. Which items are needed to light a light bulb?

A wall switch, metal plug, ignition

B metal plug, wall switch, filament

C wall switch, battery, wire

D wire, ignition, filament

---

2. What causes a bright bolt of lightning to come out of a cloud?

A The negative charges at the bottom of a cloud attract the positive charges on the ground.

B The negative charges at the top of a cloud attract the positive charges on the ground.

C The positive charges at the bottom of a cloud attract the negative charges on the ground.

D The positive charges at the top of the cloud attract the negative charges at the bottom of the cloud.

Name ____________________ Date ____________________

3. Which cup contains the liquid that has the most thermal energy?

> **Tip**
> Remember the definition of *thermal energy*.

A

B

C

D

Name ______________________ Date ______________________

4. Compare the parallel and series circuits.

Tip
Think about the differences between parallel and series circuits.

Series Circuit

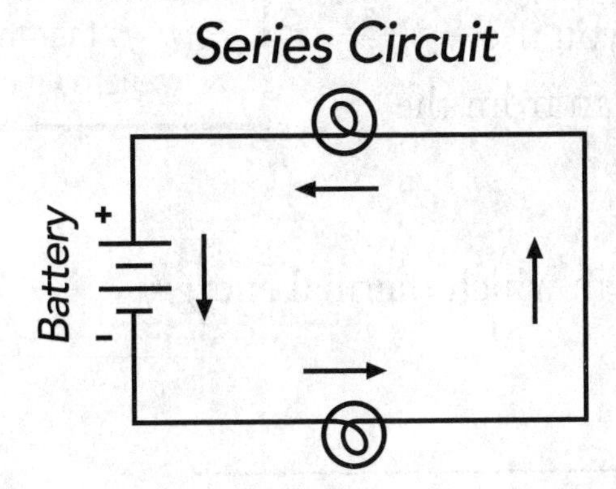

Parallel Circuit

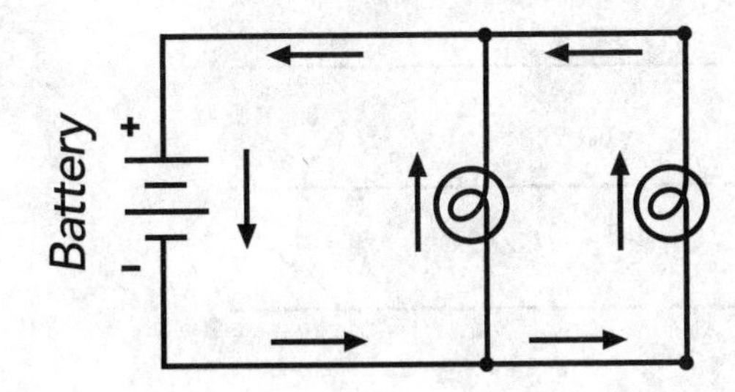

Why is a burned-out light bulb on a parallel circuit easier to identify than on a series circuit?

____

____

____

____

____

Name ______________________ Date ______________________

5. You bake a pizza on a metal tray in the oven. When it is done, you step back from the heat when you open the oven door. Then you use potholders to remove the hot metal pan from the oven.

Tip
Think about the different ways that thermal energy moves.

• Which terms best describe the ways in which thermal energy is transferred in this example?

____

____

____

____

____

____

____

• Explain why a potholder is used to remove the hot metal pan?

____

____

____

____

Name ______________________________

Date ______________________________

Chapter 16 Practice Set

Tip
Recall how to calculate speed from time and distance.

1. It takes a driver 30 minutes to drive 25 km. What speed is the driver traveling at, if he is traveling at a steady rate?

A 0.8 km per hour

B 12.5 km per hour

C 50 km per hour

D 750 km per hour

---

2. Which changes the velocity of a moving object?

A The object travels at the same speed.

B The object travels in the same direction.

C The object changes frame of reference.

D The object changes direction.

Name ______________________ Date ______________________

3. Which is an example of an unbalanced force?

> **Tip**
> Study each example, and consider the forces acting against it.

A 

B 

C 

D 

Name ____________________ Date ____________________

4. Look at the picture of the skydiver.

Tip
Think about how balanced forces will affect the skydiver.

How do different forces act on the skydiver? How will opening his parachute affect his descent?

____

____

____

____

Name ______________________ Date ______________________

5. Fish swim at different speeds as shown in the bar graph.

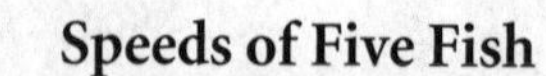

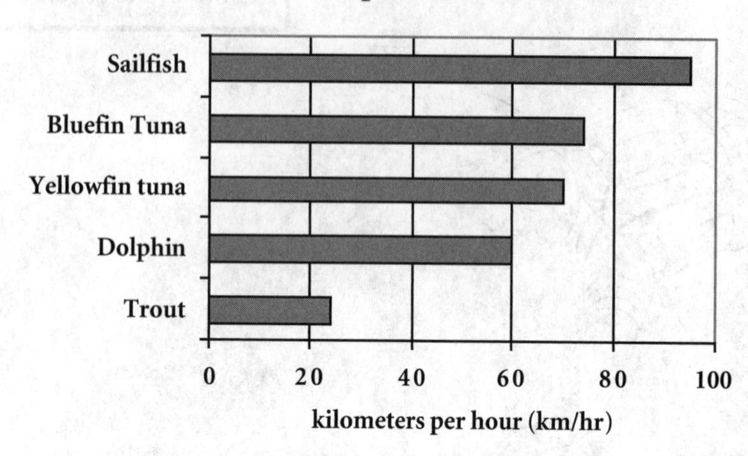

Tip
Notice how far each of the fish swims in one hour.

- If humans travel at 8 kilometers per hour, what is the difference in speed between the rate at which a human travels and the rate at which the sailfish travels?

____

____

____

- If all the fish could keep up their speed for two hours, how far would the dolphin swim in that time? Show your work.

____

____

____

Name ____________________________

Date ____________________________

1. Which simple machines is a wedge?

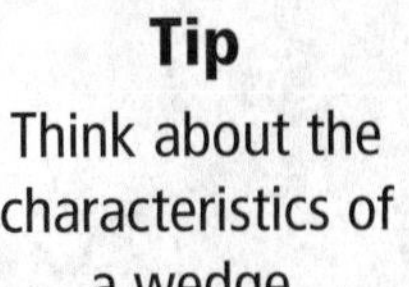

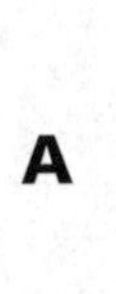

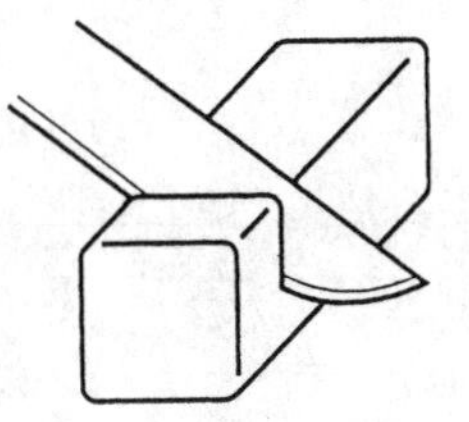

C 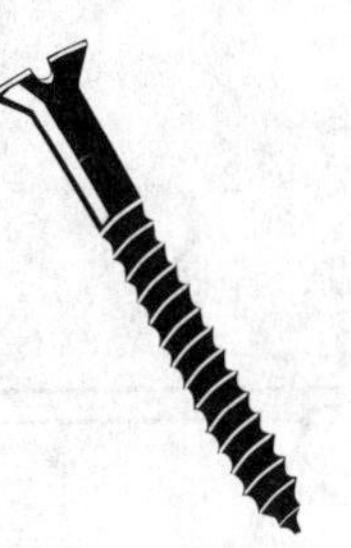

D 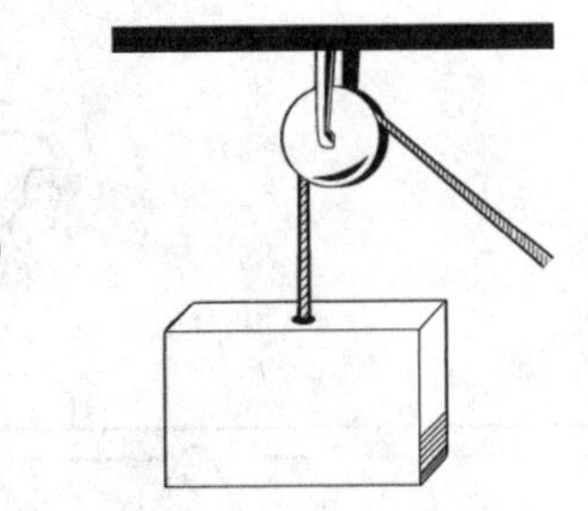

---

2. Which tool is an example of a lever?

A 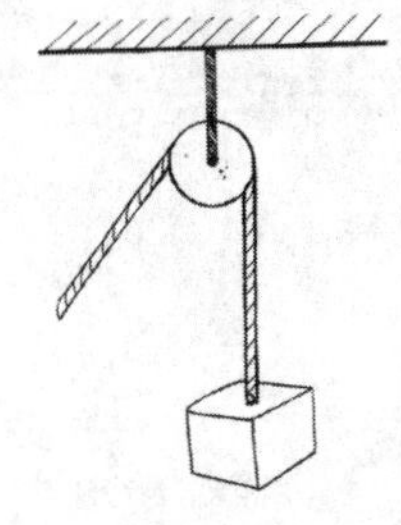

C 

B 

D 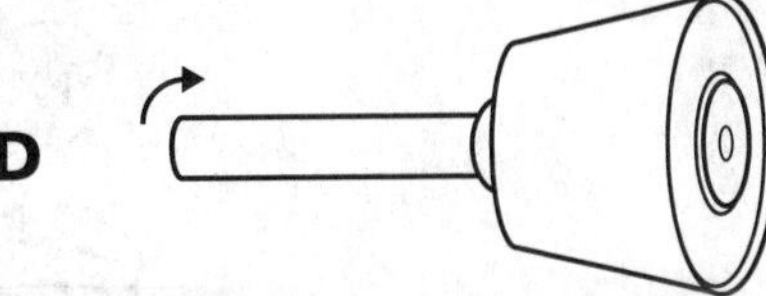

Name ______________________ Date ______________________

3. In which of the seesaws is the girl able to lift the heaviest partner?

Tip
Remember the formula for work.

A

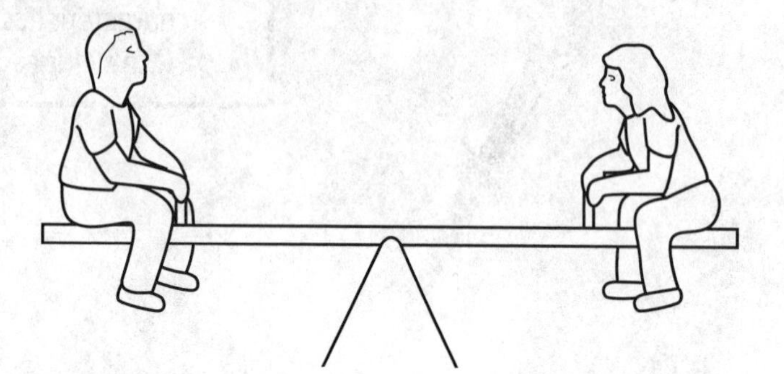

B

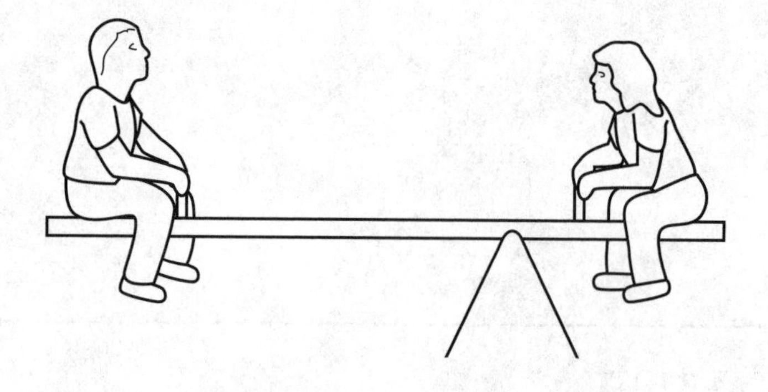

C

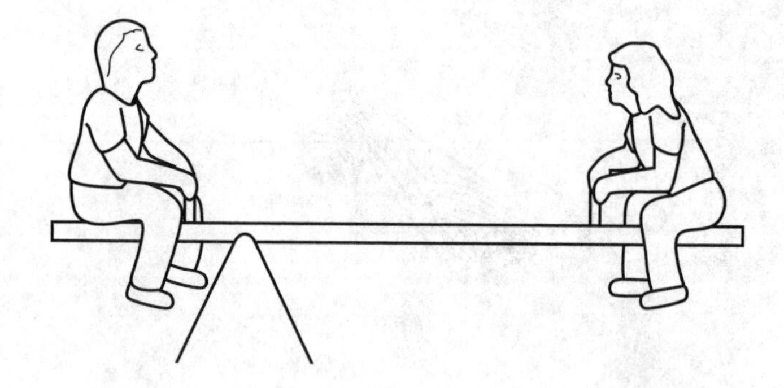

D

Name ________________ Date ________________

4. During gym class, you held a 7-pound medicine ball over your head for 1 minute. Did you do work? Why or why not?

> **Tip**
> Recall the definition of *work*.

____

____

____

____

Name ______________________ Date ______________________

5. Look at the picture. Suppose you want to move the package up the ramp to your porch.

> **Tip**
> Recall how to calculate how much work is done.

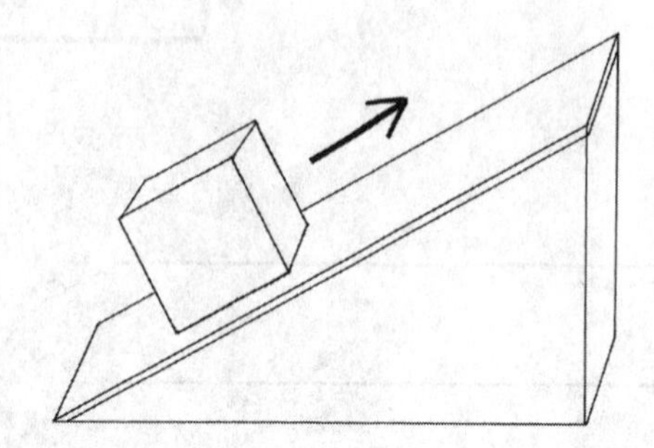

- How would you find out the amount of work required to do this job?

____

____

____

- Now suppose that you lift the box up onto the porch, without using the ramp. How would the amount of work compare?

____

____

Name ______________________________

Date ______________________________

1. The diagram shows different positions of Earth as it orbits the sun.

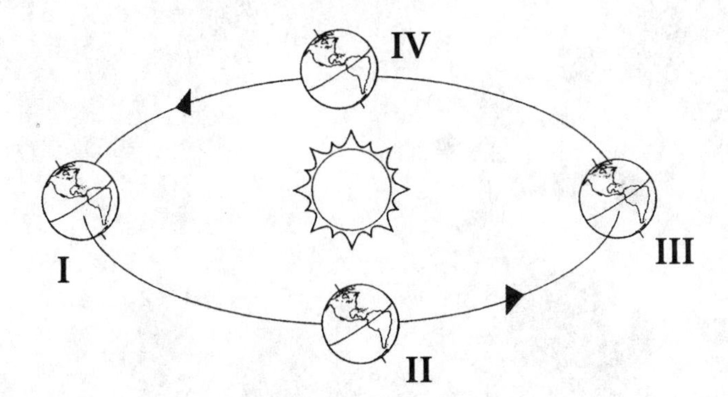

Which season is the Northern Hemisphere experiencing at point III?

A spring

B summer

C winter

D fall

5.9.6 A (2)

---

2. Before pouring your breakfast orange juice, you notice that there is pulp sitting at the bottom of the pitcher. You stir the juice. Which word describes the orange juice immediately after you stir it?

A It is a suspension.

B It is a colloid.

C It is an emulsion.

D It is an alloy.

5.6.6 A (3)

Name ______________________________

Date ______________________________

3. The diagrams below show structures found in the human body.

A 

C 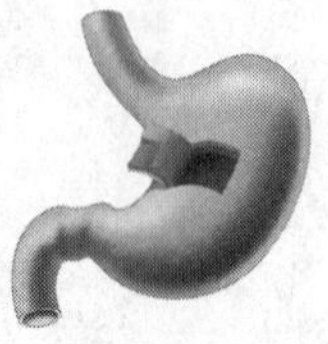

B 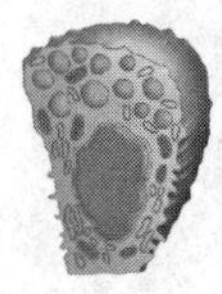

D 

Which sequence orders the structures from least complex to most complex?

A A, B, C, D

B D, C, B, A

C B, D, C, A

D C, B, A, D

5.5.6 A (1)

---

4. Today's fastest rockets can fly at about 32,000 kph. Earth's nearest star, Alpha centauri, is about 39 trillion kilometers away. About how many years would it take to travel from Earth to Alpha centauri?

A 14,246 years

B 142.46×10^{12}

C 1,248,000 years

D $1{,}248{,}000 \times 10^{12}$ years

5.9.6 B (1)

5. This diagram shows Earth's layers.

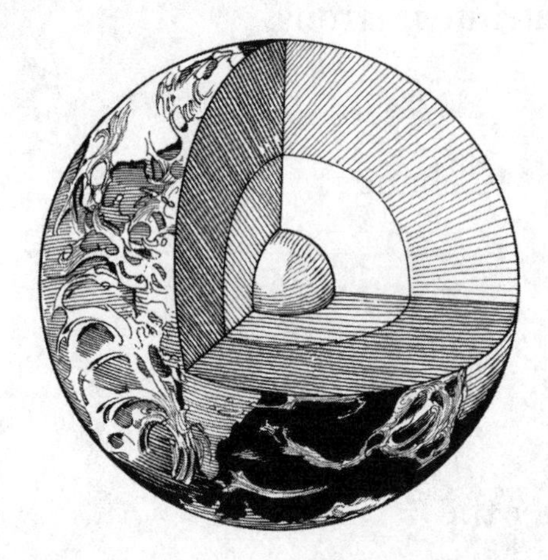

Which layer is composed of iron and nickel?

A crust

B mantle

C asthenosphere

D core

5.8.6 A

6. How do an ecosystem's biotic parts help the abiotic parts?

A Plant roots anchor soil and split rocks to make new soil.

B Bees pollinate flowers to help reproduction.

C The climate prevents insects from eating plants.

D Fungi decompose dead organisms providing nutrients for the soil.

5.10.6 A (2)

7. Two tennis balls have a weak gravitational force, but the gravitational force between Earth and a satellite is strong. Why does this occur?

A All masses have inertia, or a resistance to change in motion.

B Gravitation works through empty space to pull objects toward one another.

C The closer two objects are, the stronger the gravitational force.

D The more massive the attracting objects are, the stronger the gravitational force.

5.9.6 B (2)

---

8. A student sets up an experiment to compare the temperature of water. She wraps Beaker A with white paper and Beaker B with black paper. She fills both beakers with water at the same temperature. She sets the beakers in sunlight. After 3 hours, she records the temperature. Beaker A is 42°C and Beaker B is 46°C. What causes the temperature difference?

A More thermal energy from the sun is absorbed by the black paper.

B White paper absorbs more thermal energy.

C The white paper and the black paper absorb equal amounts of energy.

D The black paper absorbs less thermal energy from the sun.

5.7.6 B (1)

9. Study the landform shown below.

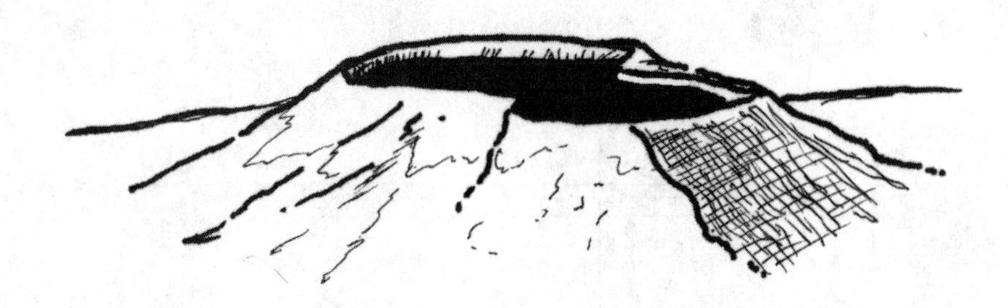

Which is the main type of rock that forms beneath and near this type of landform?

A sedimentary

B metamorphic

C igneous

D composite

5.8.6 C (1)

---

10. Ms. Jamison placed equal amounts of four unknown powders on plastic plates. She then added one drop of lemon juice to each powder, and waited to see if a chemical reaction took place. What is the independent variable in this experiment?

A the amount of lemon juice

B the different powders

C the plastic plates

D the room temperature

5.6.6 B (2)

Name ______________________________

Date ______________________________

11. The diagram shows the water cycle on Earth.

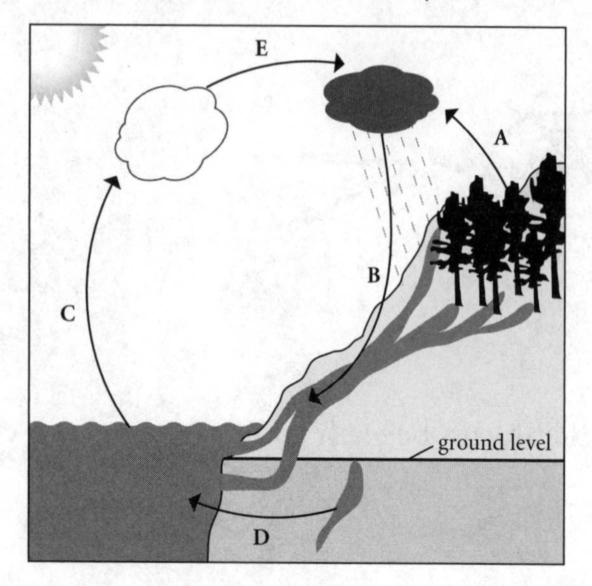

Which term best describes the process occurring at point C?

A transpiration

B evaporation

C condensation

D respiration

5.8.6 B (2)

12. In an energy pyramid, which organism would be found at the lowest point on the pyramid?

A coyote

B rabbits

C sunflowers

D soil

5.10.6 A (1)

Name ______________________________

Date ______________________________

13. Which type of information do the contour lines on topographic maps provide?

A altitude

B depth

C distance

D elevation

5.8.6 D (1)

---

14. During a walk in the forest, you observed an organism with the characteristics listed in the following table.

Characteristics		
	Yes	No
multi-celled	✓	
cell wall	✓	
nucleus	✓	
takes in food	✓	
can move		✓

How would this organism best be classified?

A plant

B animal

C protist

D fungus

5.5.6 B (1)

Name ____________________

Date ____________________

15. You design an experiment to test how height affects the distance a toy car travels. You plan to change the height of the track by placing books under one end of the track. You will have a meter stick to measure the distance traveled and height. Which of the tables shown below would be the best to record your data on?

A

Comparing Height and Distance Traveled			
Distance Traveled			
Number of books			
Height			
Speed			

B

Comparing Height and Distance Traveled			
Number of books	Height	Distance Traveled	Speed
1			
2			
3			

C

Comparing Height and Distance Traveled			
Height	Number of books	Distance Traveled	Speed

D

Comparing Height and Distance Traveled			
Height	Number of books	Speed	Distance Traveled
5 cm			
10 cm			
15 cm			

5.7.6 A (2)

Name ___________________________

Date ___________________________

16. Which contains most of Earth's fresh water?

A oceans

B rivers

C glaciers

D estuaries

5.8.6 B (1)

---

17. Study the part of the periodic table of elements.

What does the number 12 represent?

A the number of protons and neutrons in the atom's nucleus

B the number of protons in a magnesium atom's nucleus

C the number of electrons in a magnesium atom's outer shell

D the number of neutrons in a magnesium atom's nucleus

5.6.6 A (1)

Name ______________________

Date ______________________

18. Which is a way in which plants reproduce sexually?

A leaf cuttings

B tubers

C seeds

D runners

5.5.6 C (1)

---

19. The graph shows the number of extinct organisms by year.

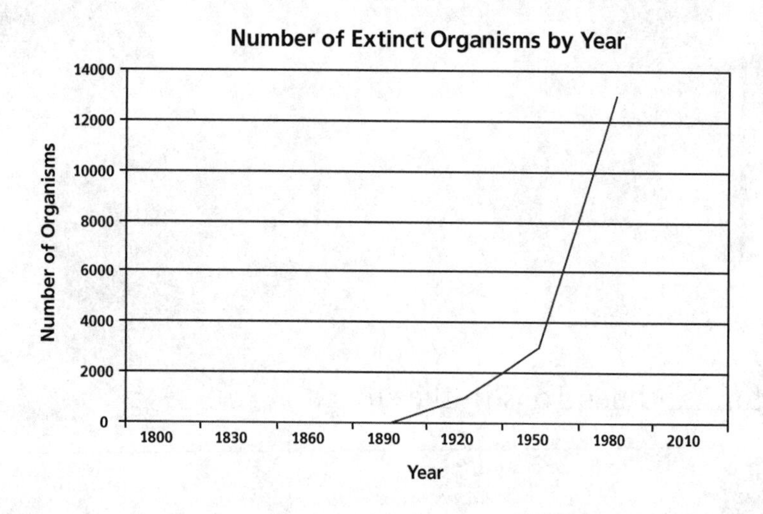

What do you predict will happen to the rate of species extinction in 2010?

A The number of extinctions will stay the same.

B The number of extinctions will decrease.

C The number of extinctions will continue to increase at a rapid rate.

D The number of extinctions will decline dramatically.

5.10.6 B (1)

Name ______________________

Date ______________________

20. Which correctly illustrates the flow of energy through a food chain?

A green algae → snail → fish → shark

B hawk → rabbit → grass → fungi

C caterpillar → robin → fungi → coyote

D snake → deer → trees → mountain lion

5.10.6 A (1)

---

21. On which object will Earth have the greatest gravitational pull?

A bowling ball

B tennis ball

C ping-pong ball

D soccer ball

5.7.6 A (3)

22. You want to find out if adding salt to water changes the temperature at which it freezes. You place 50 g salt in a graduated cylinder and add 0.5 L water. You then pour different amounts of the solution into each of four beakers and place them in the freezer. Which statement describes the freezing point of the liquids in the beakers?

A The beaker with the most salt water has the lowest freezing point.

B The beaker with the most salt water had the highest freezing point.

C The beaker with the least salt water has the lowest freezing point.

D The freezing point is the same for the water in each of the beakers.

5.6.6 A (4)

23. The graph shows a breakdown of energy consumption in the average household.

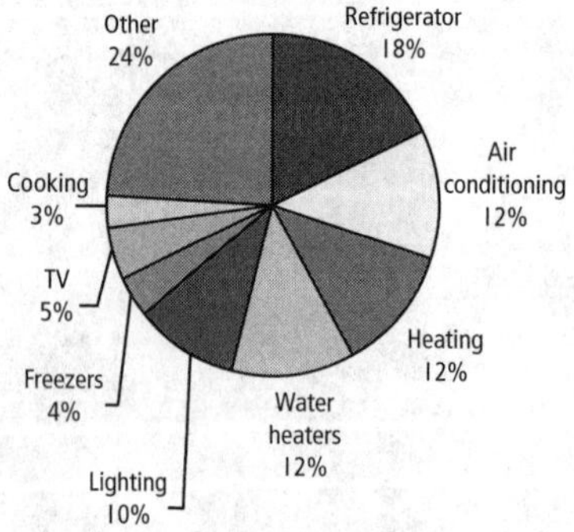

Other 24%
Refrigerator 18%
Air conditioning 12%
Heating 12%
Water heaters 12%
Lighting 10%
Freezers 4%
TV 5%
Cooking 3%

Which action would improve energy efficiency?

A Decrease the temperature setting for the air conditioner.

B Increase the temperature setting on the water heater.

C Turn up the thermostat that controls the furnace.

D Regularly vacuum the coils of the refrigerator.

5.10.6 B (2)

24. The table shows the pH of some common household items.

Item	PH
milk	6
lemon juice	2
milk of magnesia	10
drain cleaner	14
potatoes	5
orange juice	3
egg white	8
ammonia	11
carbonated soft drink	3
rainwater	5

Which item is the strongest base?

A drain cleaner

B ammonia

C lemon juice

D milk of magnesia

5.6.6 A (4)

Name ______________________________

Date ______________________________

25. Which substance has the highest temperature?

A

B

C

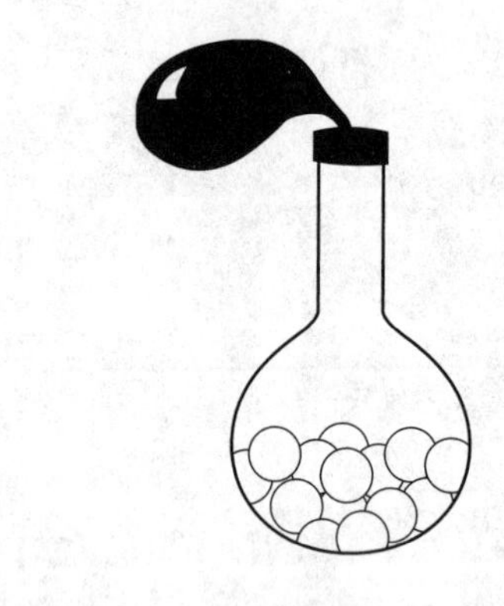

D

5.7.6 B (1)

Name ______________________

Date ______________________

26. Study the table.

Element	Occurs in Nature	State at Room Temperature	Atomic Number	Atomic Mass	Metal or Non-metal
Hydrogen	Yes	gas	1	1.01	non-metal
Nitrogen	Yes	gas	7	14.01	non-metal
Bromine	Yes	liquid	35	79.90	non-metal
Silver	Yes	solid	47	107.87	metal
Radon	Yes	gas	86	222	non-metal
Plutonium	No	solid	54	244	non-metal

Which element is least similar to the other elements in the table?

A Bromine

B Silver

C Plutonium

D Hydrogen

5.6.6 A (1)

Name ______________________

Date ______________________

27. Look at the animal and study the information in the table.

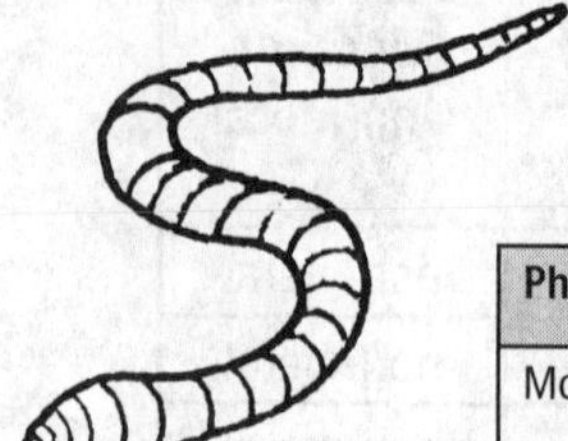

Phylum	Main Characteristics	Examples
Mollusca	soft-bodied animals with 3-part body plan; many have shells	snails, squids, clams
Nematoda	tiny, unsegmented worms; tube-shaped body; no cilia	pinworms, hook worms
Annelida	segmented worms; bristles on each segment	leeches, earthworms
Arthropoda	segmented bodies with jointed legs in pairs; many have wings	spiders, lobsters, butterflies

Which phylum does the animal belong to?

A Annelida

B Arthropoda

C Nematoda

D Mollusca

5.5.6 B (1)

---

28. Which structure in an animal cell contains chromosomes?

A cytoplasm

B vacuole

C nucleus

D mitochondria

5.5.6 A (2)

Name ______________________

Date ______________________

SCIENCE OPEN-RESPONSE

29. Biologists studied the animal populations in an area. They noticed that the lynx and rabbit populations decreased during a drought.

• In your answer document, provide a possible explanation for the decrease in the lynx population.

• A year later, the biologists noticed an increase in the rabbit population. In your answer document, identify one possible explanation for this increase.

5.10.6 A (1)

30. Throughout the night and year, the stars when viewed from Earth change position in the sky.

• In your answer document, write how the apparent nightly movement can be explained.

• In your answer document, write how the apparent yearly movement of stars can be explained.

5.9.6.C (1)

Name ______________________

Date ______________________

SCIENCE OPEN-RESPONSE

31. The United States uses more water per person than any other country. You have decided to encourage your family to help conserve this resource.

- In your answer document, describe how you might convince your family that water conservation is important.

- In your answer document, describe two behaviors your family could practice to conserve water.

5.10.6 B (2)

32. Throughout the year, the planets of the solar system appear in different parts of the night sky. When compared to the stars, the planets appear to wander around the sky.

- In your answer document, describe the shape of the planets' orbits.

- In your answer document, tell how the apparent wandering of the planets can be explained.

5.9.6 C (2)

Name ______________________________

Date ______________________________

Answer Document **Page 1**

1	Ⓐ	Ⓑ	Ⓒ	Ⓓ
2	Ⓐ	Ⓑ	Ⓒ	Ⓓ
3	Ⓐ	Ⓑ	Ⓒ	Ⓓ
4	Ⓐ	Ⓑ	Ⓒ	Ⓓ

5	Ⓐ	Ⓑ	Ⓒ	Ⓓ
6	Ⓐ	Ⓑ	Ⓒ	Ⓓ
7	Ⓐ	Ⓑ	Ⓒ	Ⓓ
8	Ⓐ	Ⓑ	Ⓒ	Ⓓ

9	Ⓐ	Ⓑ	Ⓒ	Ⓓ
10	Ⓐ	Ⓑ	Ⓒ	Ⓓ
11	Ⓐ	Ⓑ	Ⓒ	Ⓓ
12	Ⓐ	Ⓑ	Ⓒ	Ⓓ

13	Ⓐ	Ⓑ	Ⓒ	Ⓓ
14	Ⓐ	Ⓑ	Ⓒ	Ⓓ
15	Ⓐ	Ⓑ	Ⓒ	Ⓓ
16	Ⓐ	Ⓑ	Ⓒ	Ⓓ

17	Ⓐ	Ⓑ	Ⓒ	Ⓓ
18	Ⓐ	Ⓑ	Ⓒ	Ⓓ
19	Ⓐ	Ⓑ	Ⓒ	Ⓓ
20	Ⓐ	Ⓑ	Ⓒ	Ⓓ

21	Ⓐ	Ⓑ	Ⓒ	Ⓓ
22	Ⓐ	Ⓑ	Ⓒ	Ⓓ
23	Ⓐ	Ⓑ	Ⓒ	Ⓓ
24	Ⓐ	Ⓑ	Ⓒ	Ⓓ

25	Ⓐ	Ⓑ	Ⓒ	Ⓓ
26	Ⓐ	Ⓑ	Ⓒ	Ⓓ
27	Ⓐ	Ⓑ	Ⓒ	Ⓓ
28	Ⓐ	Ⓑ	Ⓒ	Ⓓ

Name ______________________________

Date ______________________________

Answer Document Page 2

Open-Response Item 29 (page 89)—4 points

- ____

____

____

____

____

____

____

- ____

____

____

____

____

____

____

Name ______________________________

Date ______________________________

Answer Document Page 3

Open-Response Item 30 (page 89)—4 points

- ____

____

____

____

____

____

____

- ____

____

____

____

____

____

____

____

Name ______________________________

Date ______________________________

Answer Document Page 4

Open-Response Item 31 (page 90)—4 points

•

•

Name ______________________________

Date ______________________________

Answer Document Page 5

Open-Response Item 32 (page 90)—4 points

- ______________________________

- ______________________________

Name ______________________

Date ______________________

1. Tara wanted to keep her canned drink cold in her lunchbox. She wrapped the can in foil, but by lunch, her drink was warm. Why was foil a poor choice to keep a canned drink cold?

A Metal is a good conductor of thermal energy.

B Metal is a poor conductor of thermal energy.

C The metal speeds up convection.

D The metal is a good insulator.

5.7.6 B (1)

---

2. Look at the map of the world and its oceans.

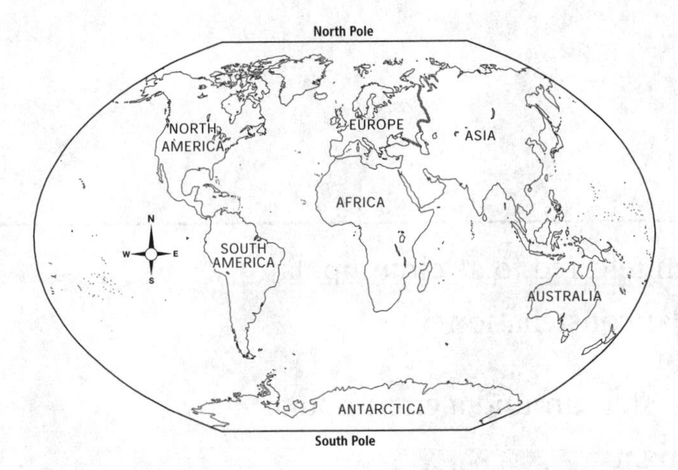

Why are ocean currents on the west coast of continents usually colder than those on the east coasts?

A The water is deeper on the west coasts of continents.

B These currents begin near the equator and flow outward.

C These currents begin near the poles where it is colder.

D The water is shallower on the west coasts of continents.

5.8.6 B (1)

Name ______________________

Date ______________________

Practice Test 2

3. The graph shows how whale populations were affected by commercial hunting.

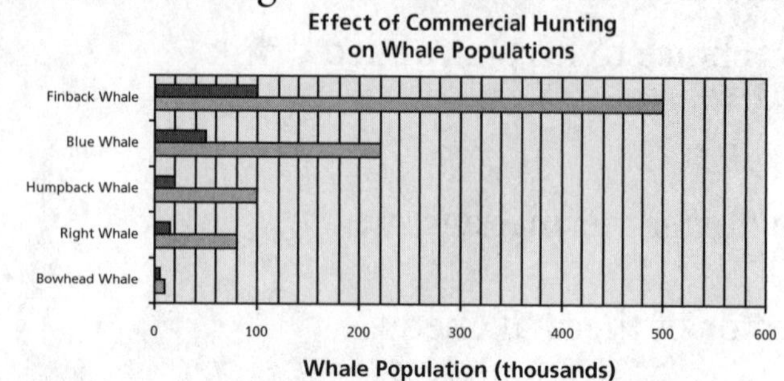

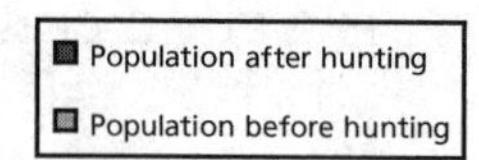

Which whale population decreased the most when humans hunted them?

A humpback whale

B blue whale

C right whale

D finback whale

5.10.6 B (1)

---

4. Certain constellations are often referred to as circumpolar. Which describes a circumpolar constellation?

A The north star appears to circle around a circumpolar constellation during the night.

B During the night, a circumpolar constellation appears to move in a circle around the pole star.

C A circumpolar constellation appears to circle around the North Pole at night.

D A circumpolar constellation is only visible from the North Pole.

5.9.6 C (1)

Name ______________________________

Date ______________________________

5. The diagram shows a map of Earth.

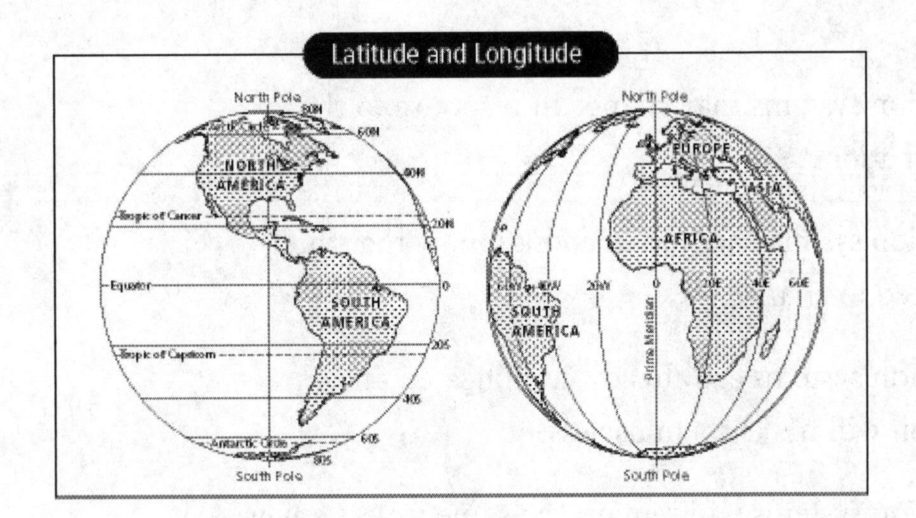

What information do lines of latitude on a map provide?

A distance east or west of the Prime Meridian

B depth of ocean water at that point

C elevation of land areas at that point

D distance north or south of the equator

5.8.6 D (1)

6. Which shows a correct order for a food chain?

A green algae → shark → fish → snail

B grass → rabbit → hawk → fungi

C caterpillar → coyote → robin → fungi

D trees → mountain lion → snake → deer

5.10.6 A (1)

Name ______________________

Date ______________________

7. What might happen to classification systems if scientists discover yet more information about the structure of living things?

A Classification systems may change in response to the new information.

B Classification systems are broad enough now that they will not need to change.

C Classification systems are already complicated and more information will make them useless.

D Classification systems will remain the same unless a new kingdom of species is found.

5.5.6 B (1)

---

8. The diagram shows the water cycle.

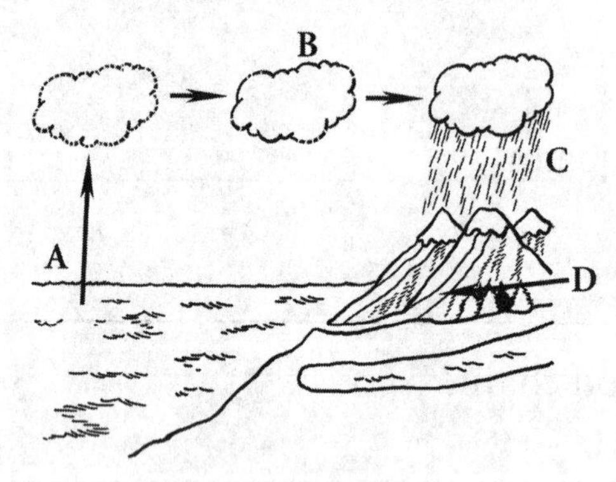

What process is occurring at point B in the diagram?

A precipitation

B evaporation

C condensation

D transpiration

5.8.6 B (2)

9. The diagram shows a phase of the moon as seen in New Jersey.

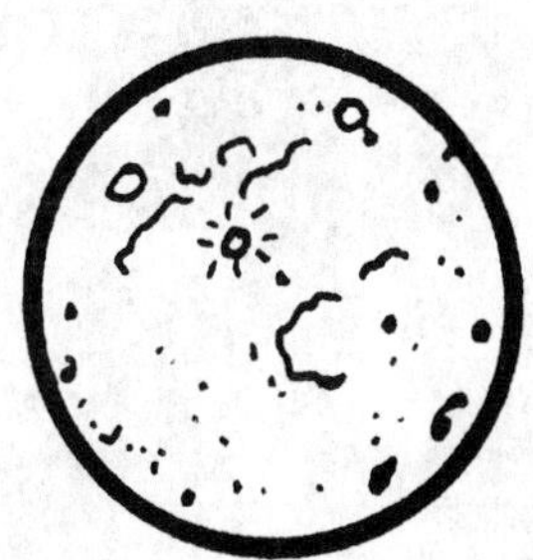

About how long will it take before this phase of the moon is seen in New Jersey again?

A 1 week

B 1 month

C 1 year

D 1 decade

5.9.6 A (1)

---

10. Your class has just finished studying the human circulatory system. You decide to conduct an experiment to find out if your heartbeat increases before, during, and after exercise. Which step would you do last in your experiment?

A Collect the materials you need to conduct the experiment.

B Count and record a friend's pulse before any exercise.

C Draw a conclusion from the results of your data.

D Count and record a friend's pulse after one minute of exercise.

5.5.6 A (1)

Name ______________________________

Date ______________________________

11. The diagram below shows a plant cell.

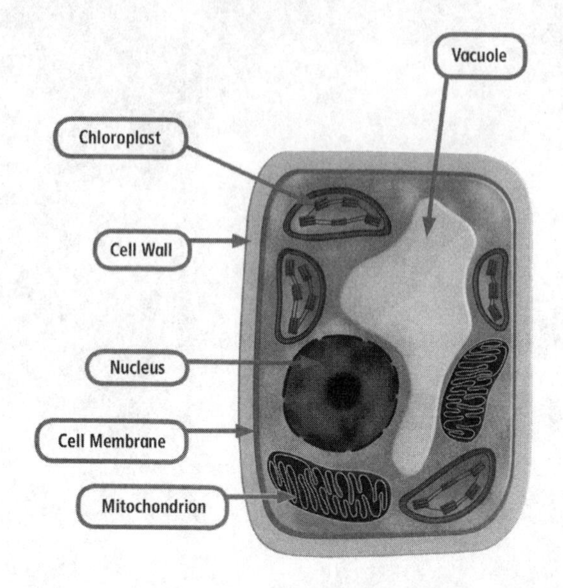

In the cell, which structure produces food?

A chloroplast

B nucleus

C mitochondria

D runners

5.5.6 A (2)

---

12. Which is a way you could conserve resources?

A turning the computer off when you leave the room

B using plastic bags at the grocery store

C driving instead of walking

D throwing away aluminum cans

5.10.6 B (2)

Name ______________________________

Date ______________________________

13. Which is a method of asexual reproduction in plants?

A germination

B tubers

C seeds

D pollination

5.5.6 C (1)

---

14. A modified rock cycle is shown below.

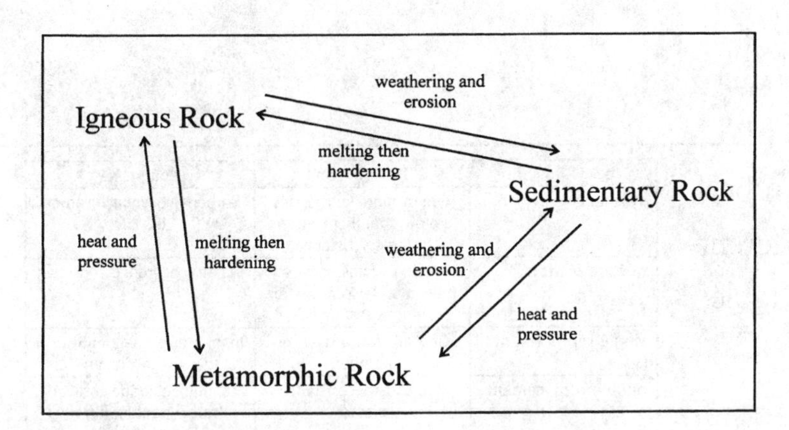

What can cause a metamorphic rock to form?

A increases in temperature and pressure

B erosion and transportation

C melting and recrystallization

D freezing and thawing

5.8.6 C (1)

Name ______________________________

Date ______________________________

15. You want to find out if newspaper or plastic foam is a better insulator. You start your investigation by filling two glass jars with hot water. Of the steps listed below, which step should be the last step in your investigation?

A Wrap one jar with newspaper and the other jar with plastic foam.

B Record the water temperature in each jar after one hour.

C Record the starting temperature of the water in each jar.

D Cover one jar with black paper and the other jar with white paper.

5.7.6 B (1)

---

16. Chemical reactions can be classified into four different types as shown in the table.

Type of Reaction	Description	Example
synthesis	Two or more substances are put together to form a single substance.	carbon + oxygen = carbon dioxide
decomposition	A single substance is broken up to form two or more substances.	water = hydrogen + oxygen
single replacement	One substance replaces another.	magnesium + lead nitrate = lead + magnesium nitrate
double replacement	Two substances switch places.	sodium chloride + silver nitrate = sodium nitrate + silver chloride

What kind of reaction occurs when the brown gas nitrogen dioxide is heated, as shown in the equation?

$$\text{Nitrogen dioxide} \xrightarrow[\text{Heat}]{} \text{nitrogen monoxide} + \text{oxygen}$$

A synthesis

B decomposition

C single replacement

D double replacement

5.6.6 B (1)

Name ______________________________

Date ______________________________

17. Which two objects will experience the strongest gravitational force?

A a satellite orbiting Earth

B a person sitting at a desk

C a tennis racquet hitting a tennis ball

D a ladybug crawling on a leaf

5.7.6 A (3)

---

18. The following is a dichotomous key used to identify a rock sample.

Step	Characteristic	Identification
1a	crystals present	go to 2
1b	crystals not present	go to 3
2a	crystals many colors	go to 4
2b	crystals all one color	Go to 4.
3a	very soft; often contains fossils	sedimentary (limestone)
3b	very hard layers	metamorphic (slate)
4a	crystals small and difficult to see	igneous, extrusive (pumice)
4b	crystals large and easy to see	igneous, intrusive (granite)

Your teacher shows you a rock sample that is made up of easy-to-see, large crystals. The crystals are of different colors including black, gray, tan, and white. What kind of rock is it?

A limestone

B slate

C pumice

D granite

5.6.6 A (1)

19. The United States uses more water per person than any other country. What can your family do to help conserve water?

A Tell your family that people in the United States use more water per person than any other country.

B Remind your family that drinking water comes from surface sources, such as rivers and lakes, and collects underground.

C Point out to your relatives that water demand is greater than available water.

D Ask your family members to take showers instead of baths.

5.10.6 B (2)

20. Study the groups of items.

Which property of the items was most likely used to group the items?

A shape of substance

B mass of substance

C chemical properties

D state of matter

5.6.6 A (4)

Name ______________________________

Date ______________________________

21. Which kingdom of organism is described by the following characteristics: cell walls, multi-celled, make their own food, nucleus, and cannot move?

A Animalia

B Fungi

C Plantae

D Protista

5.5.6 (B) 1

22. You want to find out how speed is related to distance. You design an experiment using the following setup.

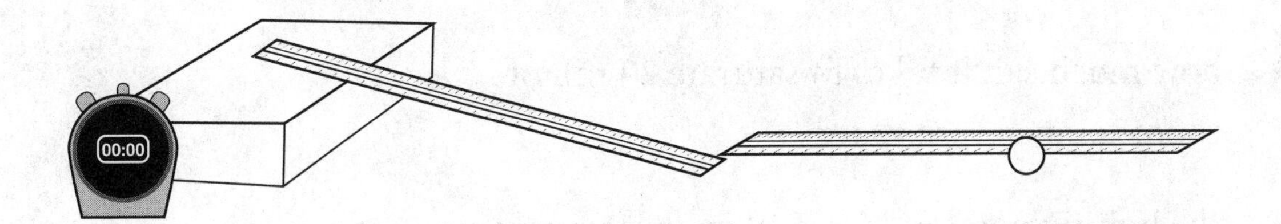

You plan to measure the speed a marble can travel down the plane. What is the independent variable in your experiment?

A the height of the block under the ruler

B the size of the marble

C the starting point of the marble

D the length of the ruler

5.7.6 A (1)

Name ______________________

Date ______________________

23. What do the colors of the stars tell scientists about the stars?

A brightness

B age

C mass

D temperature

5.9.6 D

---

24. Your teacher has 20 mL of room-temperature water in a beaker. She adds 1 gram of salt, stirs it, and repeats the process until she sees salt particles at the bottom of the beaker. What is your teacher most likely trying to determine through this experiment?

A how much salt it takes to saturate 20 mL of room-temperature water

B the boiling point of 20 mL of room-temperature water and 1 gram of salt

C the chemical properties of mixing salt with water

D the density of salt water, like the water found in the world's oceans

5.6.6 A (3)

25. Study the information in the formula.

$$\text{carbon dioxide} + \text{water} \xrightarrow[\text{chlorophyll}]{\text{sunlight} +} \text{sugar} + \text{oxygen}$$

What process does the formula describe?

A respiration

B reproduction

C photosynthesis

D phototropism

5.10.6 A (1)

26. Sam wants to know if the mass of water changes when it is frozen and then allowed to melt. Of the steps listed below, which should be the first step in Sam's experiment?

A Determine the mass of the water and its container before freezing.

B Determine the mass of the water and its container after the ice has melted.

C Determine the mass of the water after freezing.

D Determine the mass of the container and then the container filled with water.

5.6.6 A (4)

Name ______________________

Date ______________________

27. Distances in space are often measured in light years. What is a light year?

A the distance that light travels in one year

B the shortest time that it takes to travel from Earth to the moon

C the amount of time it takes for sunlight to travel from the sun to Earth

D the distance between the sun and Earth

5.9.6 B (1)

---

28. Study the graph showing the relationship between a car's speed and the rate at which it can stop.

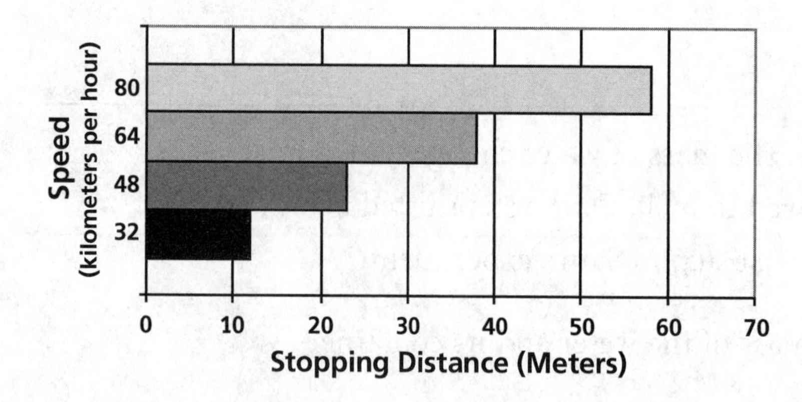

At which speed does it take the car the longest time to stop?

A 32 km/hr

B 64 km/hr

C 48 km/hr

D 80 km/hr

5.7.6 A (2)

Name ______________________________

Date ______________________________

SCIENCE OPEN-RESPONSE

29. Adam goes on a hike with his family. He sees few trees, but many plants with long narrow leaves and thick fibrous roots. The diagram shows these adaptations.

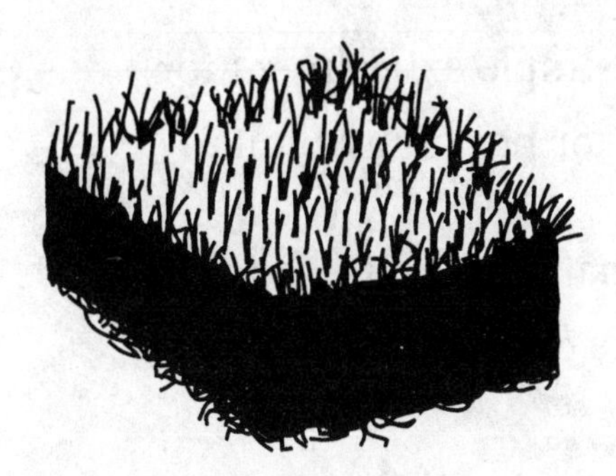

- In your answer document, explain how these adaptations help the plants survive in their environment.

- In your answer document, identify how the grasses might adapt if there was a season of particularly low rainfall.

5.10.6 A (1)

Name ___________________________

Date ___________________________

SCIENCE OPEN-RESPONSE

30. Mary built a container to collect "gray water" from her house. Gray water is waste water from showers, clothes washers, and sinks. It can be used to wash cars and water plants.

• In your answer document, describe a simple experiment for Mary to use to test if gray water is safe for her houseplants.

• In your answer document, identify what data Mary should collect, and explain how she can use the data to draw conclusions.

5.10.6 B (2)

Name ______________________________

Date ______________________________

SCIENCE OPEN-RESPONSE

31. Earth is made of four distinct layers.

• In your answer document, list the layers that make up Earth from the innermost layer to the outer layer.

• In your answer document, describe how a hard-boiled egg can be used as a model for Earth.

5.8.6.A

Name ______________________________

Date ______________________________

SCIENCE OPEN-RESPONSE

32. Gravity is a force that acts on objects without touching them. Gravity is a non-contact force.

- In your answer document, name the two factors that affect gravitational attraction between two objects.

- In your answer document, explain how these two factors affect the gravitational attraction between the objects.

5.9.6 B (2)

Name ______________________________

Date ______________________________

Answer Document **Page 1**

1 Ⓐ Ⓑ Ⓒ Ⓓ
2 Ⓐ Ⓑ Ⓒ Ⓓ
3 Ⓐ Ⓑ Ⓒ Ⓓ
4 Ⓐ Ⓑ Ⓒ Ⓓ

5 Ⓐ Ⓑ Ⓒ Ⓓ
6 Ⓐ Ⓑ Ⓒ Ⓓ
7 Ⓐ Ⓑ Ⓒ Ⓓ
8 Ⓐ Ⓑ Ⓒ Ⓓ

9 Ⓐ Ⓑ Ⓒ Ⓓ
10 Ⓐ Ⓑ Ⓒ Ⓓ
11 Ⓐ Ⓑ Ⓒ Ⓓ
12 Ⓐ Ⓑ Ⓒ Ⓓ

13 Ⓐ Ⓑ Ⓒ Ⓓ
14 Ⓐ Ⓑ Ⓒ Ⓓ
15 Ⓐ Ⓑ Ⓒ Ⓓ
16 Ⓐ Ⓑ Ⓒ Ⓓ

17 Ⓐ Ⓑ Ⓒ Ⓓ
18 Ⓐ Ⓑ Ⓒ Ⓓ
19 Ⓐ Ⓑ Ⓒ Ⓓ
20 Ⓐ Ⓑ Ⓒ Ⓓ

21 Ⓐ Ⓑ Ⓒ Ⓓ
22 Ⓐ Ⓑ Ⓒ Ⓓ
23 Ⓐ Ⓑ Ⓒ Ⓓ
24 Ⓐ Ⓑ Ⓒ Ⓓ

25 Ⓐ Ⓑ Ⓒ Ⓓ
26 Ⓐ Ⓑ Ⓒ Ⓓ
27 Ⓐ Ⓑ Ⓒ Ⓓ
28 Ⓐ Ⓑ Ⓒ Ⓓ

Name ___________________________

Date ___________________________

Answer Document Page 2

Open-Response Item 29 (page 111)—4 points

•

•

Name ______________________________

Date ______________________________

Answer Document Page 3

Open-Response Item 30 (page 112)—4 points

- ______________________________

- ______________________________

Name ______________________________

Date ______________________________

Answer Document **Page 4**

Open-Response Item 31 (page 113)—4 points

•

•

Name ______________________________

Date ______________________________

Answer Document Page 5

Open-Response Item 32 (page 114)—4 points

-

-

Name ______________________

Date ______________________

1. An eagle eats a snake that ate a grasshopper that ate some grass. In this food chain, what is the eagle?

A an herbivore

B a consumer

C a producer

D a decomposer

5.10.6 A

---

2. Scientists have identified an object orbiting the sun. The object has a year equal to about 24 Earth years. Between which two planets is the object located?

Planet	Time to Orbit the Sun in Earth Years
Mercury	88 Earth days
Venus	225 Earth days
Earth	365.25 Earth days
Mars	1.9 Earth years
Jupiter	11.9 Earth years
Saturn	29.5 Earth years
Uranus	84 Earth years
Neptune	165 Earth years
Pluto	248 Earth years

A between Earth and Mars

B between Jupiter and Saturn

C between Uranus and Neptune

D between Neptune and Pluto

5.9.6 A (1)

Name ______________________________

Date ______________________________

3. Look at the cell.

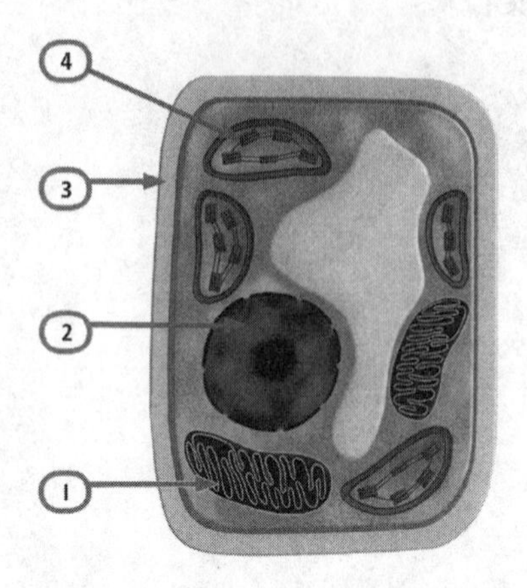

What is the name of the structure labeled 1?

A cell wall

B nucleus

C chloroplast

D mitochondrion

5.5.6 A (2)

4. Two organisms live in the same area and eat the same types of food. Which sentence best describes their relationship in their environment?

A They share the same niche within a habitat.

B They live in different habitats but share the same niche.

C They have a predator/prey relationship.

D They share the same habitat but have different niches.

5.10.6 A (1)

5. Scientists estimate the distance from Earth to stars using a concept called parallax. What is parallax?

A the change in position of a star when viewed from two different points in Earth's orbit

B the change in position of a star when viewed from two different points on Earth's surface

C the change in brightness of a star when viewed from two different places on Earth's surface

D the change in brightness of a star when viewed from two different points in Earth's orbit

5.9.6 B (1)

---

6. A fire sweeps through an oak-hickory forest, burning everything in its path. Which plants will be the first to grow back?

A oak tree seedlings

B shrubs

C water plants

D grasses

5.10.6 A (2)

7. Study the information in the table.

Water Type	Description
A	mixed waters with different salt concentrations
B	high salinity
C	fresh water
D	fresh water with low oxygen concentration

Which water type describes waters present in estuaries?

A water type A

B water type B

C water type C

D water type D

5.8.6 B (1)

8. The information below represents the distance traveled by four bicycle riders over a period of time. Which bicycle rider has the fastest speed?

A 150 km ÷ 2 hours

B 160 km ÷ 2 hours

C 80 km ÷ 2 hours

D 200 km ÷ 4 hours

5.7.6 A (1)

Name ___________________________

Date ___________________________

9. Douglas planted a bean in potting soil and added a small amount of water.

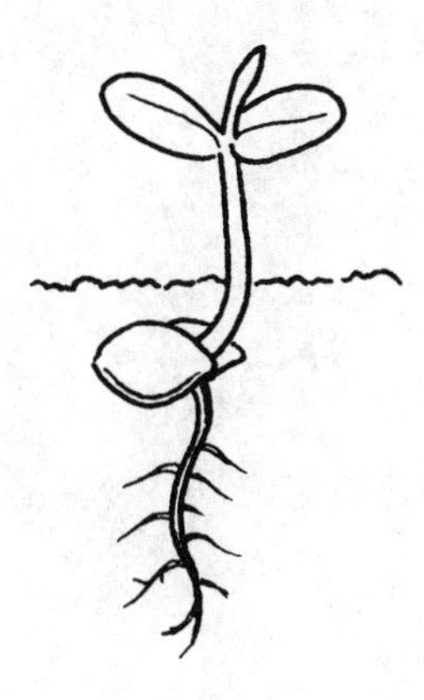

Which stage of the bean plant's life cycle is shown?

A germination

B leaf growth

C reproduction

D flower production

5.5.6 C (1)

---

10. During Earth's orbit around the sun, what occurs when the Northern Hemisphere experiences summer?

A The Northern Hemisphere is very close to the sun.

B The Northern Hemisphere is tilted away from the sun.

C The Northern Hemisphere is pointing directly at the sun.

D The Northern Hemisphere is tilted towards the sun.

5.9.6 A (2)

Name ______________________________

Date ______________________________

11. The diagram below shows the rock cycle.

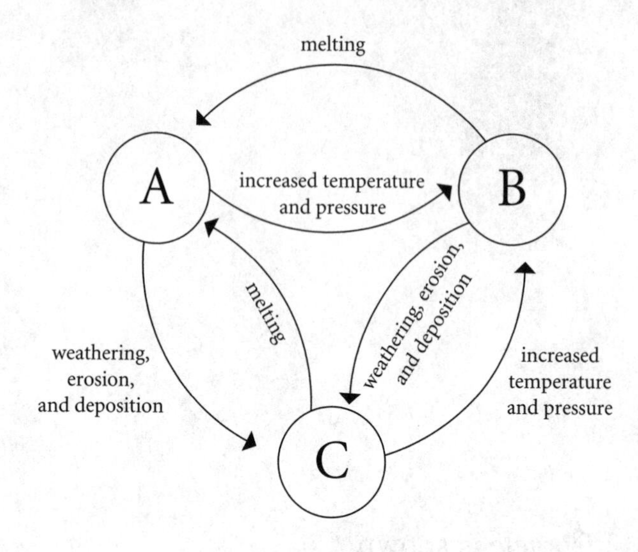

Which type of rock forms from erosion of other rocks?

A igneous

B metamorphic

C sedimentary

D magmatic

5.8.6 C (1)

---

12. A rabbit has babies. Some of the baby rabbits have white fur. Others have black fur. Some have black and white fur. What do the different colors show?

A mimicry

B adaptations

C reproduction

D variation

5.5.6 B (2)

Name ______________________________

Date ______________________________

13. The gravitational pull between Earth and the moon is responsible for the oceans' tides. Which lists the correct arrangement for a high tide or spring tide to occur?

A the sun, the moon, and Earth aligned in a row

B the sun, the moon, and Earth in scattered positions

C the sun, the moon, and Earth forming a right angle

D the sun, the moon, and Earth forming an acute angle

5.9.6 B (2)

---

14. The picture shows a map of Earth.

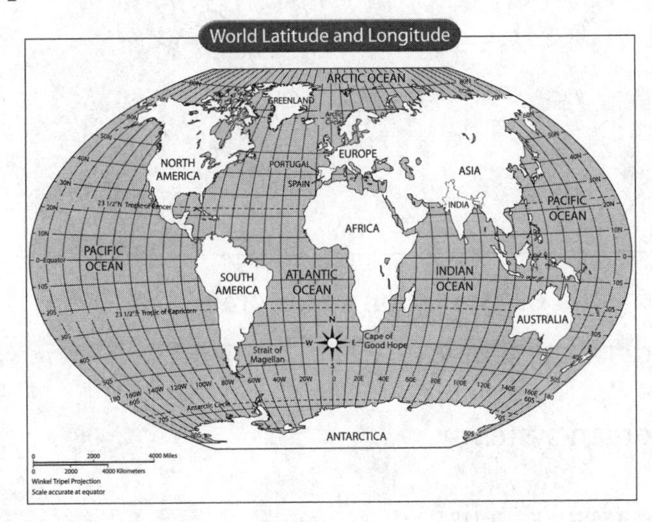

What information do lines of longitude on a map provide?

A distance north or south of the equator

B diameter of Earth at that point

C distance east or west of the Prime Meridian

D elevation of Earth at that point

5.8.6 D (1)

Name ______________________________

Date ______________________________

15. Your teacher asked you to look at the different substances found in a drop of a liquid mixture. Which instrument should you use?

A

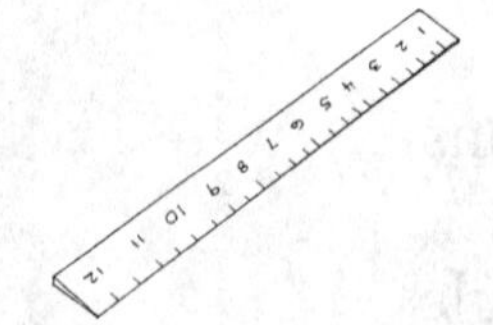

C

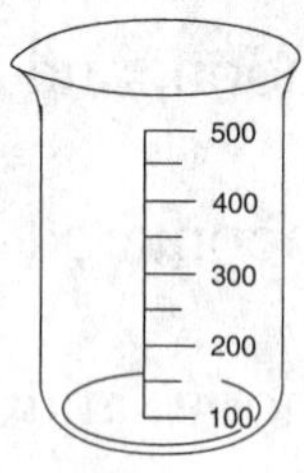

B

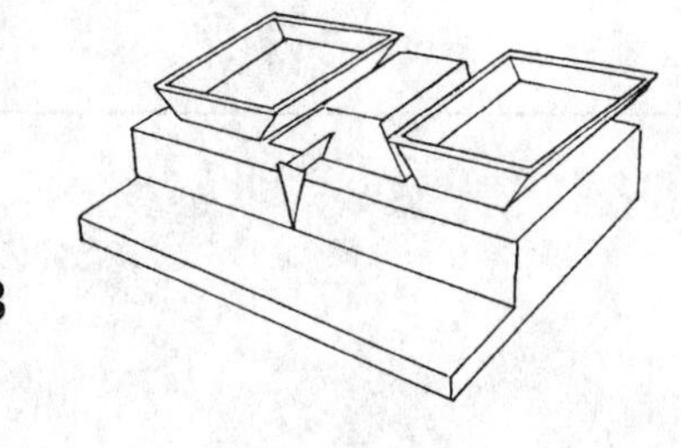

D

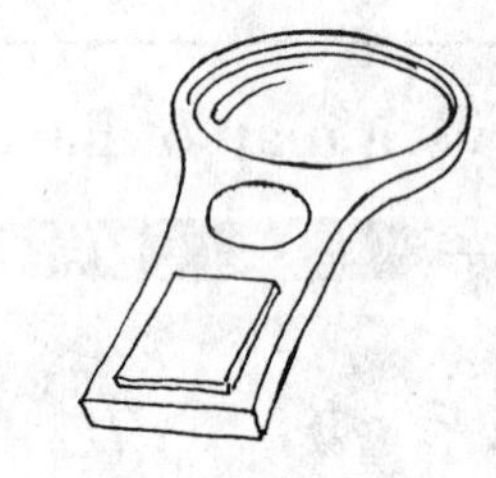

5.6.6 A (3)

---

16. Which is the correct sequence of organization in a living thing from least complex to most complex?

A cells, organs, tissues, organ systems

B organs, organ systems, tissues, cells

C tissues, organs, cells, organ systems

D cells, tissues, organs, organ systems

5.6.6 A (1)

Name ____________________

Date ____________________

17. Study the parts of the periodic table that are shaded gray.

15	16	17	18
			2 He Helium
7 N Nitrogen	8 O Oxygen	9 F Fluorine	10 Ne Neon
15 P Phosphorus	16 S Sulfur	17 Cl Chlorine	18 Ar Argon
33 As Arsenic	34 Se Selenium	35 Br Bromine	36 Kr Krypton

What do the shaded elements have in common?

A They are all part of the same period.

B They are all solids at room temperature.

C They are all members of the same family.

D They do not occur in nature.

5.6.6 A (1)

18. You design an experiment to see how friction affects movement of a toy car. Which materials would **most likely** help you with this experiment?

A thermometer, beaker, and balance scale

B sandpaper, aluminum foil, and a carpet square

C piece of wood, timer, ruler

D balance scale, block of wood, plastic bag

5.7.6 A (2)

Name ______________________

Date ______________________

19. The graph shows how an average home uses electricity.

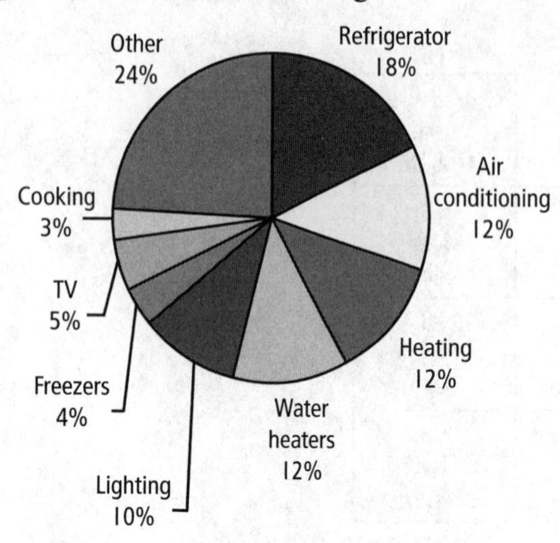

If you wanted to save money on electricity, which of the following should you do?

A Set the air conditioner to a higher temperature.

B Watch more TV in the dark.

C Use the microwave to cook instead of the stove.

D Turn the refrigerator to a colder temperature.

5.10.6 B (2)

---

20. An experiment tests how the height of a ramp affects the distance a toy car will travel. What variable changes in the experiment?

A the size of the car

B the surface of the ramp

C the length of the ramp

D the height of the ramp

5.7.6 A (1)

Name ______________________

Date ______________________

21. There are four categories of chemical reactions, as shown.

Type of Reaction	Description	Example
Synthesis	Two or more substances are put together to form a single substance.	carbon + oxygen = carbon dioxide
Decomposition	A single substance is broken up to form two or more substances.	water = hydrogen + oxygen
Single replacement	One substance replaces another.	magnesium + lead nitrate = lead + magnesium nitrate
Double replacement	Two substances switch places.	sodium chloride + silver nitrate = sodium nitrate + silver chloride

Which reaction is an example of a double replacement reaction?

A sodium chromate + silver nitrate = sodium nitrate + silver chromate

B nitrogen dioxide = nitrogen monoxide + oxygen

C hydrogen sulfide + silver = silver sulfide + hydrogen

D sodium + chlorine = sodium chloride

5.6.6 B (1)

---

22. In a forest food chain, a bear eats both berries and fish. How would the bear be classified based on the type of food it eats?

A an herbivore

B a producer

C a carnivore

D an omnivore

5.10.6 A (1)

Name ___________________________

Date ___________________________

23. The diagram below shows the setup of an experiment.

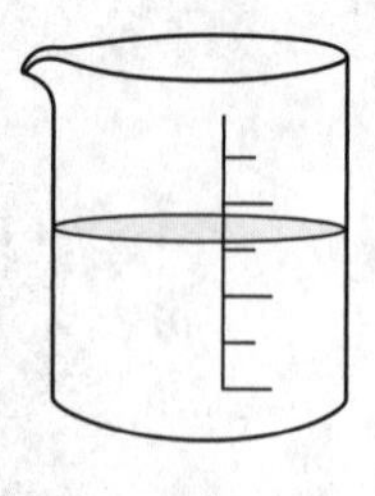

A
water +
1 teaspoon
baking soda

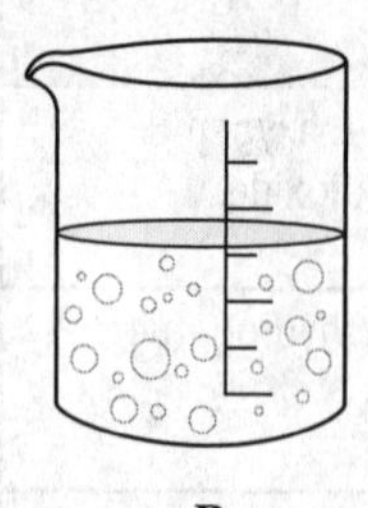

B
vinegar +
1 teaspoon
baking soda

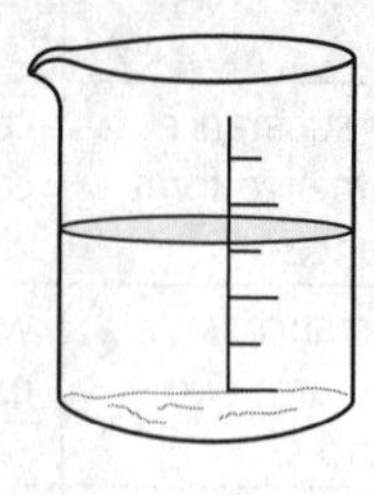

C
cooking oil +
1 teaspoon
baking soda

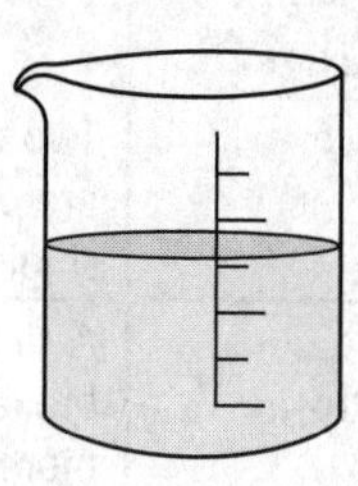

D
milk +
1 teaspoon
baking soda

In which jar will a chemical reaction take place?

A A

B B

C C

D D

5.6.6 B (1)

---

24. Your class tested how some liquids chemically change rocks. You placed pieces of the same rock in three different liquids. What is the independent variable in this experiment?

A the liquid in the beakers

B the size of the pieces of rock

C the time the rocks are in the liquids

D the volume of liquid in the beakers

5.6.6 B (1)

25. Use the dichotomous key to help you classify a solar system.

Step	Characteristic	Identification
1a	huge, curved arms	Go to 2.
1b	irregular shape	Go to 3.
2a	edge looks like flattened disk; old and new stars	spiral galaxy
2b	edge has bar structure	barred spiral galaxy
3a	flat, small, mostly old stars	elliptical galaxy
3b	no distinct pattern or arrangement; mostly young stars	irregular galaxy

Scientists are studying a galaxy. The galaxy does not have a clearly formed shape. It is made up mostly of old stars. Which kind of galaxy are the scientists **most likely** studying?

A elliptical galaxy

B barred spiral galaxy

C spiral galaxy

D irregular galaxy

5.9.6 D

26. You want to find out which liquids absorb the most heat in sunlight. You write the steps of an experiment.

Step 1:	Take the starting temperature of each liquid.
Step 2:	Cover each jar with plastic wrap.
Step 3:	Take the ending temperature of each liquid.
Step 4:	Place each jar of liquid in a sunny window.

What is the correct order of steps to follow in the experiment?

A 1, 2. 3, 4

B 1, 2, 4, 3

C 4, 3, 2, 1

D 2, 1, 4, 3

5.7.6 B (1)

27. Study the graph of elements in the human body.

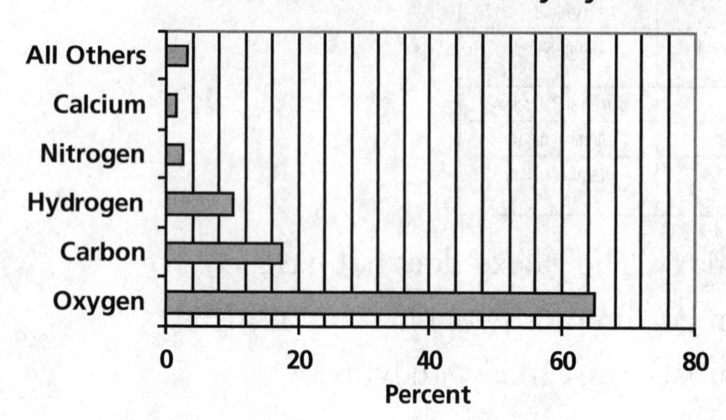

Which element is **most common** in the human body?

A oxygen

B carbon

C hydrogen

D nitrogen

5.6.6 A (1)

Name ______________________

Date ______________________

28. You want to do a simple experiment by placing butter on the pancakes.

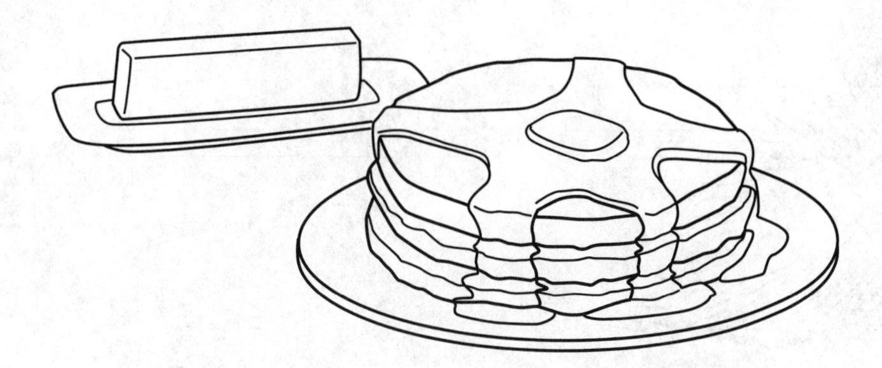

What physical change will you observe during this experiment?

A Butter melts on the pancakes.

B The temperature of the pancakes increases.

C Pancakes melt with the butter.

D The temperature of the butter decreases.

5.6.6 A (4)

Name ______________________________

Date ______________________________

SCIENCE OPEN-RESPONSE

29. Study the characteristics of the animal shown.

- In your answer document, explain how this animal's body type helps it survive in its environment.

- The animal has teeth that are especially good for gnawing tough substances. In your answer document, tell what type of environment you think it lives in.

5.10.6 A (1)

Name ______________________________

Date ______________________________

SCIENCE OPEN-RESPONSE

30. Study the food chain below.

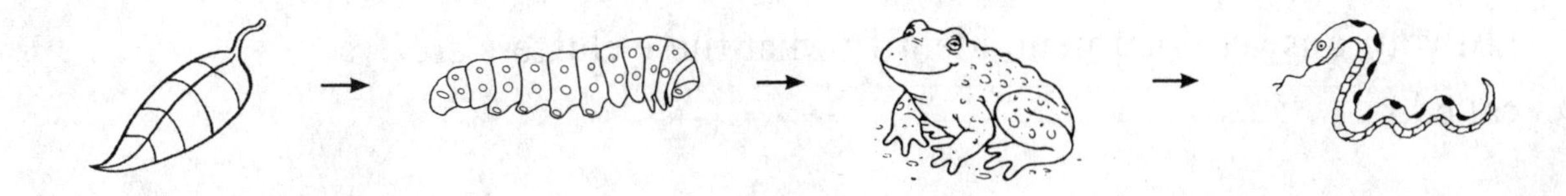

• In your answer document, identify the correct sequence of the organisms in the food chain.

• In your answer document, explain what would happen if the fertilizer from local farms entered the water supply, killing tadpoles and removing frogs from the food chain.

5.10.6 B (1)

Name ______________________________

Date ______________________________

SCIENCE OPEN-RESPONSE

31. Throughout history, people have recognized specific clusters of stars in the sky.

• In your answer document, identify what these clusters are called.

• In your answer document, explain why some of these clusters are visible only during certain times of the year.

5.9.6 C (2)

---

32. It takes Earth about 365 days to orbit the sun. It takes Halley's comet about 76 years to orbit the sun.

• In your answer document, identify the number of Earth years one year is equal to on Halley's comet.

• In your answer document, use the length of Halley's comet's year to describe the size of its orbit compared to Earth and one other object in our solar system.

5.9.6 D

Name ______________________________

Date ______________________________

Answer Document **Page 1**

1 Ⓐ Ⓑ Ⓒ Ⓓ
2 Ⓐ Ⓑ Ⓒ Ⓓ
3 Ⓐ Ⓑ Ⓒ Ⓓ
4 Ⓐ Ⓑ Ⓒ Ⓓ

5 Ⓐ Ⓑ Ⓒ Ⓓ
6 Ⓐ Ⓑ Ⓒ Ⓓ
7 Ⓐ Ⓑ Ⓒ Ⓓ
8 Ⓐ Ⓑ Ⓒ Ⓓ

9 Ⓐ Ⓑ Ⓒ Ⓓ
10 Ⓐ Ⓑ Ⓒ Ⓓ
11 Ⓐ Ⓑ Ⓒ Ⓓ
12 Ⓐ Ⓑ Ⓒ Ⓓ

13 Ⓐ Ⓑ Ⓒ Ⓓ
14 Ⓐ Ⓑ Ⓒ Ⓓ
15 Ⓐ Ⓑ Ⓒ Ⓓ
16 Ⓐ Ⓑ Ⓒ Ⓓ

17 Ⓐ Ⓑ Ⓒ Ⓓ
18 Ⓐ Ⓑ Ⓒ Ⓓ
19 Ⓐ Ⓑ Ⓒ Ⓓ
20 Ⓐ Ⓑ Ⓒ Ⓓ

21 Ⓐ Ⓑ Ⓒ Ⓓ
22 Ⓐ Ⓑ Ⓒ Ⓓ
23 Ⓐ Ⓑ Ⓒ Ⓓ
24 Ⓐ Ⓑ Ⓒ Ⓓ

25 Ⓐ Ⓑ Ⓒ Ⓓ
26 Ⓐ Ⓑ Ⓒ Ⓓ
27 Ⓑ Ⓑ Ⓒ Ⓓ
28 Ⓐ Ⓑ Ⓒ Ⓓ

Name ______________________________

Date ______________________________

Answer Document Page 2

Open-Response Item 29 (page 96)—4 points

- ____

____

____

____

____

____

____

- ____

____

____

____

____

____

____

Name ______________________________

Date ______________________________

Answer Document Page 3

Open-Response Item 30 (page 97)—4 points

- ____

- ____

Name ______________________________

Date ______________________________

Answer Document Page 4

Open-Response Item 31 (page 98)—4 points

•

•

Name ______________________________

Date ______________________________

Answer Document **Page 5**

Open-Response Item 32 (page 98)—4 points

•

•